I0494335

WELLNESS
for
ELEPHANTS

Proceedings of the Jacksonville Workshop

Edited By
MEGAN C. MORRIS, VALERIE D. SEGURA
DEBRA L. FORTHMAN *and* TERRY L. MAPLE

Fernandina Beach, FL, Red Leaf Press 2019

Dedicated to Dr. Michael Hutchins, who provided essential guidance to the AZA elephant program during a challenging time in its history. Mike's ideas and his personal courage made AZA zoos more responsive to the needs of elephants, and we will always remember his contributions to elephant welfare and conservation.

TABLE OF CONTENTS

List of Contributing Authors

Chapter 1

Terry Maple
Director of Wildlife Wellness, Jacksonville Zoo and Gardens
Email: Terrylmaple@msn.com

Megan Morris
Formerly, Jacksonville Zoo and Gardens
Current Affiliation: Animal Wellness Coordinator, Fresno Chaffee Zoo
Email: MMorris@fresnochaffeezoozoo.org

Kaylin Tennant
Graduate Research Associate, Cleveland MetroParks Zoo
kst@clevelandmetroparks.com

Valerie Segura
Curator of Wildlife Wellness and Behavioral Science, Jacksonville Zoo and Gardens
Email: Segurav@jacksonvillezoo.org

Chapter 2

S.K. McGuinness, FRGS

Chapter 3

Bob Lee
Elephant Curator and Manager, Oregon Zoo
Bob.Lee@oregonzoo.org

Sharon Stuart Glaeser
Research Associate, PhD Candidate, Oregon Zoo
Sharon.Glaeser@oregonzoo.org

Chapter 4

Erin Ivory
Curator of Mammals
IUCN Asian Elephant Specialist Member, North Carolina Erin.
ivory@nczoo.org

Corinne J. Kendall, PhD
Curator of Conservation and Research,
North Carolina Zoo
corinne.kendall@nczoo.org

Guy Lichty
Retired, North Carolina Zoo
guylichty@yahoo.com

Chapter 5

Chris D. Peterson
Associate Curator of Animals/Elephant Manager,
Cleveland Metroparks Zoo
cdp@clevelandmetroparks.com

Bonnie A. Baird, PhD
Formerly Cleveland MetroParks Zoo
Current Affiliation: Animal Welfare Scientist, Woodland Park Zoo
Bonnie.Baird@zoo.org

Chapter 6

Megan Wilson, PhD
Lecturer; Coordinator, Undergraduate Research,
Georgia State University
mwilson72@gsu.edu

Chapter 7

Lauren Highfill, Ph.D.
Professor of Psychology and Animal Studies,
Eckerd College
HighfiLE@eckerd.edu

Otto Fad
Animal Welfare & Behavior Specialist, Precision Behavior
Ofad@precisionbehavior.com

Jessica Spencer
Busch Gardens Tampa Bay
jmspence26@gmail.com

Chapter 8

Joseph Soltis
Animal Welfare Science Manager
Animals, Science and Environment
Disney's Animal Kingdom®
Joseph.Soltis@disney.com

David A. Orban
Formerly, Disney's Animal Kingdom®
Current affiliation: Animal Excellence Manager,
Cincinnati Zoo & Botanical Gardens
David.Orban@cincinnatizoo.org

Kimberly J. Adams
Research Associate
Animals, Science and Environment
Disney's Animal Kingdom®
Kimberly.J.Adams@disney.com

Jll Mellen
Director of Science and Education
Animals, Science and Environment
Disney's Animal Kingdom®
(retired)

Lori Perkins
Formerly, Disney's Animal Kingdom®
Current affiliation: Deputy Director,
Birmingham Zoo
lperkins@birminghamzoo.com

Chapter 9

Margaret Whittaker
Creative Animal Behavior Solutions, Active Environments, Inc,
Oakland Zoo
Indu22@earthlink.net

Chapter 10

S.G. Friedman
Professor Emeritus, Department of Psychology, Utah State
University
Behavior Works, LLC
sgfriedman@icloud.com

Chapter 11

Kate Evans
Founder & Director, Elephants for Africa
kate@elephantsforafrica.org

Chapter 12

Lara Metrione
Research Associate,
South East Zoo Alliance for Reproduction and Conservation
Lara.Metrione@sezarc.com

Patrick Flora
Formerly Birmingham Zoo, Inc
Patrickflr50@yahoo.com

William Foster
Formerly Birmingham Zoo, Inc

Linda Penfold
Director, South-East Zoo Alliance for Reproduction & Conservation
linda.penfold@sezarc.com

Chapter 13

Elizabeth (Betsy) Herrelko
Smithsonian's National Zoo
HerrelkoE@si.edu

Tony Barthel
Smithsonian's National Zoo
BarthelT@si.edu

Chapter 14

Nevin Lash, ASLA
Principal, Ursa International
nevin@ursainternational.org

Chapter 15

Meredith J. Bashaw,
Franklin & Marshall College
mbashaw@fandm.edu

Chapter 16

Gordon M. Burghardt
Departments of Psychology and Ecology & Evolutionary Biology,
University of Tennessee
gburghar@utk.edu

Chapter 17

Terry Maple
Director of Wildlife Wellness, Jacksonville Zoo and Gardens
Email: Terrylmaple@msn.com

Jeff Sawyer, ASLA, LEED AP BD+C
Senior Associate,
CLR Design inc
jsawyer@clrdesign.com
Gary Lee
CLR Design
glee@clrdesign.com

Julia Hanuliakova
Design Principal
Zoo Design, Inc.
jhanuliakova@zoodesign.co

Dan Maloney
Deputy Zoo Director/Animal Care and Conservation, Jacksonville
Zoo and Gardens
Maloneyd@jacksonvillezoo.org

FOREWORD

Wellness for Elephants is based on a workshop hosted by the Jacksonville Zoo and Gardens in March 2016. We are grateful to the authors of these collected papers who worked so hard to share their ideas and their findings with other professionals committed to superior facilities and care for African and Asian elephants. This meeting was organized by the Jacksonville Zoo Wellness Department under the direction of Dr. Terry L. Maple. The three Jacksonville coeditors were joined by independent consultant, Dr. Debra Forthman, who offered her expertise as a scholar and editor in mammalian behavior to help produce this book. The meeting was supported by the entire Jacksonville Zoo and Gardens staff, who provided support for the work and social sessions that combined to create a memorable gathering.

We anticipate future workshops of this kind, blending wellness research and services and providing new ideas and approaches to the exhibition of wild animals. In this volume, two contributions provide a glimpse of our future focus: Burghardt's chapter on reptiles and Bashaw's chapter on giraffe. We believe that the wellness construct is inclusive in its application to virtually all species that are managed by zoos and aquariums. As our wellness team concluded, wellness is an expansion of animal welfare and promises to deliver new opportunities for zoo and aquarium animals to thrive in human care.

We offer this collection of papers to advance wellness as a viable approach to animal welfare and encourage our colleagues in the zoo world to apply and test this new standard in the operation of their facilities. We welcome those who read this book to visit our zoo to see wellness-inspired design and management in action.

Tony Vecchio
CEO, Jacksonville Zoo and Gardens

Chapter 1

DESIGNING AND MANAGING FOR WELLNESS

Terry L. Maple, Megan C. Morris, Kaylin S. Tennant, and
Valerie D. Segura
Jacksonville Zoo and Gardens/University of North Florida

The global commitment to institutional animal welfare is a trans-formative step for accredited zoos and aquariums. Comprehensive reforms have rapidly elevated zoological standards and practices to unprecedented levels of excellence. The recent publication of a strategic vision of animal welfare by scholars working on behalf of the World Association of Zoos and Aquariums (WAZA) makes it abundantly clear that conservation and welfare are now regarded as essentially mutually synergistic and of equal importance (Mellor, Hunt, & Gusset, 2015). Zoo and aquarium leaders must now build on this momentum by creating organizational structures and committing sufficient financial resources to fully implement these priorities. By significantly upgrading the quality of life for animals living in the world's zoos and aquariums, we will likely influence the willingness of our visitors and donors to continue their support for our cause. The enthusiastic support of the public is necessary to continue the revolutionary change that is taking our institutions from good to great. This giving paradigm is elegant in its simplicity: design, build, and operate facilities where the animals happily thrive to the delight of all who are educated and inspired by enlightened exhibition.

Scientific zoo biologists have generated many publications focused on zoo and aquarium populations, but the foundation of zoo animal welfare is derived from decades of systematic research on domestic animals utilized in agriculture and biomedicine (Brambell, 1965). Historically, animal welfare in biomedical and agricultural settings was connected to overall physical health and, therefore, overseen by veterinarians. The first concern of administrators and regulators was the issue of farm and biomedical subjects living in conditions that produced suffering. Eventually, both governmental and private sector specialists promulgated new regulations which led to higher standards and better practices to reform these unacceptable conditions. With the improvement of these standards and practices, zoos and aquariums took notice. Modern zoos and aquariums are regulated by governments and by their own set of accreditation standards (Norton, Hutchins, Stevens, & Maple, 1995; Maple, 2016). The world's accredited zoos and aquariums are easily differentiated from roadside attractions that lack a credible program of evaluation.

A broader approach to animal welfare took hold when psychologists became active contributors to the field. Long before the animal welfare movement was launched, pioneers in primate behavior research such as Robert M. Yerkes (1925, 1943) and Harry F. Harlow (1932, 1971) and their many students and collaborators carried out basic research that identified the psychological needs of monkeys and apes. The construct *psychological well-being* was introduced in 1988 by two former students of Harlow: Melinda Novak and Steve Suomi. A number of published collections subsequently examined animal welfare for primates and other species (Erwin, Maple, & Mitchell, 1979; Segal, 1989; Novak & Petto, 1991). Hal Markowitz, a psychologist who contributed significantly to the zoo animal welfare movement, was the scholar who launched innovations at the Portland (now Oregon) and San Francisco Zoos (Markowitz, 1982). His ideas were also introduced at many other zoos and aquatic facilities around the world. Markowitz is still an iconic figure in the animal welfare movement. His published work should be required reading for all serious students of zoo animal welfare.

JACKSONVILLE'S WELLNESS WORKSHOP

In March 2016, leaders and scholars at the Jacksonville Zoo and Gardens organized and hosted a Wellness for Elephants workshop to introduce "wellness" as an appropriate construct for defining the full scope of animal welfare for elephants in managed care. The Jacksonville team elected to focus this meeting on elephants, although two additional talks examined the utility of wellness for other zoo taxa. *Wellness for Elephants* is the first compendium to test the wellness construct as a new approach in design and management (Maple & Bocian, 2013; Maple & Perdue, 2013). With the publication of this organized survey of research and practice, we anticipate a body of new research as investigators discover how this construct can be usefully applied to a diversity of wildlife in human care. It is highly likely that we will convene additional meetings to investigate and discuss specialized issues in wellness design, research, and practice. Many North American zoos including the Brevard Zoo, Palm Beach Zoo, and San Francisco Zoo have utilized the wellness construct as a guiding principle in animal management. In San Francisco, wellness has become a brand signifying the zoo's strong commitment to animal welfare. From our home base in Jacksonville, we are tracking the dissemination and performance of the wellness model, and we have recently exchanged ideas with zoos in Australia and Brazil.

WELLNESS OR WELFARE?

How is wellness different from welfare? Since the discipline of animal welfare is historically connected with conditions that lead to suffering, early studies were concerned with reforms and prevention. The original goal of prevention produced animal welfare standards that were frankly minimal rather than optimal with the intent to meet regulatory standards widely applied and monitored by associations and governments. Investigators of animal welfare in zoos have asserted that most of the animals confined for agriculture, biomedicine, or exhibition no longer suffer; instead, domestic, lab, and zoo animals have learned to cope with the challenging conditions of captivity (Maple & Perdue, 2013). In Jacksonville, we have taken the strong position that while coping is better than suffering, coping is not good enough (Maple, 2016; Maple & Segura, 2018). It is far better to set standards higher, so animals are encouraged to thrive or flourish. Our operating mantra has become "fewer species, living large." Traditional zoos were expected to assemble a large and diverse collection of wildlife often at the expense of quality of life. Before landscape immersion techniques were widely adopted, small cages barely suitable for one or two animals were the norm. By contrast, in a zoo committed to wellness, the species-appropriate group size in the wild serves to guide the number of animals that should occupy an enlarged, naturalistic, and complex enclosure.

A key principle in advancing elephant wellness is to populate a herd that encourages socialization, breeding, and parenting. In nature, elephant herds are led by a mature matriarch. Bulls grow up in the herd, leave it, and periodically return to it. Bulls also associate in male herds where older bulls help to socialize younger ones (Evans, chapter 9 of this volume). A wellness-inspired zoo will always design exhibits around the psychological needs of wildlife. As zoo professionals have come to understand the needs of elephants, we have stopped the practice of chaining, begun to introduce and manage bulls, built barns with soft substrates and yards with full-immersion pools of water, activated animals through training, added contiguous trails so they could walk and run, and opened night house doors so the herd could be socially active outdoors at night. In contrast to welfare regulations, wellness is optimal and aspirational. Other than the cost and boundaries of our creativity, there is no upper limit to wellness. True believers in wellness will find a way to make it happen and expand the frontiers of animal welfare.

In business, we associate thriving with success or prosperity. In public health, thriving is associated with well-being. Those at the top tier of business and health are described as thriving. We define wellness as equivalent to a standard of optimal welfare achieved when they are behaving like wild animals (Maple & Bloomsmith, 2018; Tennant,

Segura, Morris & Maple, 2018). For example, in the new elephant habitat at the Dallas Zoo, educators placed a sign proclaiming "let elephants be elephants." On a 4-acre site where they are encouraged to exercise, some of the Dallas animals are walking more than 5 miles per day. The Dallas elephants and those living in other wellness-inspired zoo habitats such as those in Cleveland; Oakland; Omaha; San Diego; Washington, DC; and Wichita meet the definition of thriving. The day is coming when all elephants in AZA-accredited zoos will be encouraged to thrive in a new generation of exhibits.

EVIDENCE-BASED DESIGN

One of the goals of "wellness-inspired design" is to determine how to reliably activate conditions of wellness. To do so, we must be able to define and identify the components of thriving for a given species and individual conspecifics. For every species and every individual, the prototype for wellness is the animal as it behaves in its natural habitat. Zoos that are dedicated to knowing and understanding the animals that we manage are obligated to systematically study the animals in their care and gain mastery of the published literature. There is no better justification for a dedicated or collaborative research department than its vital role in the advancement of wellness. Management by the application of evidence is the benchmark of a truly empirical and ultimately ethical zoo. The application of field data was advocated long ago by Hediger (1955), who discussed it in his book *Studies on the Psychology and Behavior of Captive Animals in Zoos and Circuses:*

> In addition to the problem of the amount of space needed by an animal, the problem of its quality has long been neglected . . . we are indebted to the field biologists for the essential information. To the animal that lives in it, a territory may not be of equal value from one end to the other, but is in fact in various ways subdivided into

different localities, each association with definite functions and significance . . . there are sleeping quarters,
bathing places, food stores, etc.; in short, special inside
arrangements differing for various species. (pp. 8).

Hediger applied his knowledge of field biology in directing three Swiss
zoos in the cities of Basel, Bern, and Zurich. For example, he insisted that
zoos utilize curvilinear design as much as possible. This design principle
is evident in the famed Africa House at Zoo Zurich. He regarded rectangular buildings as essentially "unbiological," as they confined animals to
a life disadvantaged by corners (Hediger, 1970). He simulated nature by
creating substrates that were uneven rather than flat. For rhinos, hippos,
and elephants, Hediger provided rocks and trees so the animals could
rub against them. Logs were imbedded in concrete walls to provide additional naturalistic rubbing surfaces. He noted that rhinos rubbing
their horns on concrete and steel risked serious damage to their horn.
Comfort was an important feature of Hediger's type of simulated habitat. No zoo designer should attempt to build any animal exhibit without
reading Hediger's vast collection of books and papers. Just as important,
enlightened design requires a thorough review of the psychological profile of each species in the zoo. For this reason, zoo architects often work
closely with knowledgeable curators and behavioral scientists. During
the early stages of Zoo Atlanta's revitalized design process, the zoo hired
a specialist with the title "zoo biologist" whose primary responsibility
was to monitor and inform the design and construction team.

A BRIEF HISTORY OF WELLNESS

The literature of human psychology must be probed also to identify
the most important features of wellness. Wellness is ubiquitous in fitness, medical, and spiritual settings. College infirmaries have been replaced with wellness centers that emphasize prevention and holistic
physical and mental health. On the west coast in particular, wellness is a

strategy for living. In fact, people want to reach the highest possible levels of wellness, which has become a major focus of the human potential movement revealed in the words of pioneers such as A.H. Maslow. Also known as "humanistic psychology," the human potential movement concentrated on people who had reached a high state of functioning. "Self-actualization" was the ideal that Maslow developed to describe this state of being. The most intriguing aspect of self-actualization is how easily it translates from the human to the nonhuman realm. Maslow's hierarchy of needs is particularly relevant to animal welfare, and it is depicted in a graphic form in the WAZA strategic document for animal welfare (Hanuliakova, 2015. The fact that Maslow studied rhesus monkeys as H. F. Harlow's first graduate student at the University of Wisconsin may have played a role in this facile application. An additional framework for achieving one's full potential was provided in Maslow's classic book *Toward a Psychology of Being* (1962). Maple (1996) modified his ideas in the list below. For both animals and people, we can easily recognize:

- The existence of a biologically based and uniquely individual inner nature.

- A nature that can be studied and discovered.

- A nature than is neutral, pre-moral, or good.

- A nature that should be encouraged; its suppression leads to sickness.

- A nature that is strong and unmistakable.

- A nature that is always pressing for actualization.

- Overcoming obstacles results in healthy self-esteem.

The final item in this list is especially important for animals in managed care. Traditional zoos restricted opportunities for animals, feeding them on an unchanging schedule and requiring no effort to acquire food. There is no challenge in getting food at the same time in the same place each and every day. A better way to feed zoo animals is to provide food on a variable interval schedule while scattering the food throughout the enclosure. The animals don't know when the food is coming, and they must forage for the food that is provided. Labeled "contra-freeloading," psychologists have consistently demonstrated that animals prefer to work for their food (Jensen, 1963). Both Yerkes and Hediger understood that animals required an active life in the lab and the zoo. According to Hediger (1955):

> The problem of occupational therapy, of providing pastimes in the zoo . . . has been given more and more attention, to the great benefit of the animals. We are no longer content to let animals vegetate stupidly in narrow cages, but do our utmost to see that their lives are healthy and full, and as positive as possible. pp. 38

ENGINEERING OPPORTUNITIES

The first use of the term "wellness" in a publication about animal behavior appeared in a paper by Fritz and Howell (1993). This paper described the operating philosophy of the Arizona Primate Foundation, a facility largely devoted to chimpanzees. The owners, Jo and Paul Fritz, were especially concerned with the well-being of the animals, as many had been confined in substandard laboratories, zoos, and private homes. The rescued animals were given outdoor activity daily and opportunities to climb and explore. They were also provided with material for nesting in addition to enrichment objects and toys for mental stimulation. A variety of food items were spread throughout the day so the animals could search for them. All animals lived in mixed-sex groups. Their approach was clearly successful, but for some reason, the construct of wellness did not reappear in the primate literature. While wellness was a new organizing construct, the management techniques described in this important paper were not new. In his book *Almost Human*, Yerkes (1925) promoted a similar program of husbandry based on his observations of the chimpanzee colony at Quinta Palatino, Cuba. However, the Arizona Primate Foundation demonstrated leadership in putting chimpanzee welfare first on their management agenda. In the early 90s, even the world's best zoos were playing catch-up in their implementation of better practices and higher standards. Wellness ideas have been on the table for a long time, but they are not yet the norm.

Wellness-inspired exhibits for elephants and other animals must take into account the normal social structure of wild animals. For elephants, the exhibits we build must be large enough for a group of elephants. Because elephants are very large indeed, these exhibits must encompass both sufficient acreage and sufficient volumetric space. The best exhibits will include space devoted to male groups and space where males and females can safely interact. They must also provide for migration simulations that encourage elephants to walk to destinations that offer enrichment (e.g., water, food, or play). While we advocate larger exhibits for elephants, we disagree with some assessments that sanctuary-size exhibits (1,000 acres or more) should be the norm. In the wild, elephants migrate long distances to find water and food, but the typical standard for active elephants is closer to what our better zoos provide (e.g., at least 4 acres). Many of America's best elephant exhibits also feature cognitive work stations where the animals can operate manipulanda to produce rewards or solve problems. In colder climates, massive night houses are necessary as the Cologne Zoo has constructed (Figure 1), but if weather permits, elephants should always have access to the outdoors at night. To engineer environments this large, zoos must devote a significant percentage of their campus to elephants and other resident species of megafauna (see Bashaw, chapter 13 of this volume, for a discussion of the needs of giraffe in zoos). Environmental upgrades for elephants have been discussed in a number of recent publications (Kane,

Forthman, & Hancocks, 2009; Hancocks, 2009; Maple, Bloomsmith, & Martin, 2009; Maple & Perdue, 2013; Maple, 2016) indicating a growing consensus on the proper way to exhibit elephants. Needless to say, we believe elephants will thrive in managed care if zoos thoughtfully apply their knowledge of the elephant's psychological needs. Clearly, zoo designers now grasp the problem of exhibiting elephants, and their recent work indicates they are willing to go the extra mile to provide elephants with an acceptable quality of life. Of course, to achieve optimal wellness standards, zoos must be willing to obtain the extraordinary funding that is necessary to build expansive, invulnerable, innovative, safe enclosures for a large herd of large animals. One way to ensure we are on the right track is to involve field biologists with a comprehensive knowledge of elephant behavior and ecology. Kate Evans, represented in this volume, is one of the field scientists who have freely provided good advice to zoo designers and managers.

FIGURE 1. INDOOR ELEPHANT EXHIBIT AT COLOGNE (GERMANY) ZOO.

MANAGING ELEPHANTS
TO ACHIEVE WELLNESS

Wellness-inspired exhibits provide opportunities to thrive. However, thriving is best achieved when animals are activated by a motivated keeper staff, dedicated trainers, and contingencies arranged through appropriate technology. Environmental enrichment is a component of wellness. It is a subject discussed in great detail in our book *Zoo Animal Welfare* (Maple & Perdue, 2013). Keepers who introduce objects that can be manipulated by elephants must also monitor their use and continuously refresh the objects to maintain their interest. As in nature, there must be variation in managed care. We believe the involvement of dedicated or consulting behavior analysts (e.g., Maple & Segura, 2015) provides greater flexibility in the use of conditioning methods to activate elephants. For zoos that employ trainers or keepers who train, they still benefit by contact with certified behavior analysts. Such expertise can be found in the zoo profession (e.g., the zoo consultants, *Active Environments*) or in nearby universities or institutes. From our contacts with academic behavior analysts, we know there is considerable interest in working with zoo professionals to advance the welfare of animals. We hope the behavior analyst position at Jacksonville Zoo and Gardens is not a temporary fad but a promising trend. For example, staff trained in behavior analysis have been employed recently at Disney's Animal Kingdom.

Hal Markowitz (1982, 2011) provided basic information about the technical systems he advocated and deployed to enrich the lives of zoo animals. During his long career, many zoos utilized specialized technology to enrich big cats, primates, elephants, and other creatures, but the equipment often fell into disrepair and could not be salvaged by zoo staff. Markowitz's last book was written to give zookeepers the knowledge they need to design, build, and repair apparatus installed in zoo enclosures. Today, mental activation is carried out digitally with sophisticated manipulanda (often joysticks) that operate a computerized work screen. Just like teenagers glued to the tube at home, zoo animals are fascinated by the opportunity to solve problems and gain rewards. In many zoos, elephants have been presented with ropes and levers to operate feeders, deliver showers, and open doors. Some of these devices enable elephants to control lights and sounds within their exhibits. Essentially, these innovations are designed to engineer autonomy and choice. At the San Francisco Zoo, Vice President of Wellness and Animal Behavior Jason Watters tinkers with advanced technology, and he and his students do it in a lab named for Hal Markowitz. Today's wellness/welfare technology will deliver new breakthroughs that benefit wildlife in zoos. It is fitting that Watters is creating new tools and technology at the zoo where Hal Markowitz tutored a generation of zoo biologists.

THE WELLNESS AGENDA

We planned and hosted the Wellness for Elephants workshop in Jacksonville to examine the potential of the wellness construct. We asked our speakers to share their ideas and their data within the context of a wellness-operating philosophy. The papers we are sharing in this book reveal innovations taking place in the world's best zoos. These innovations include newly built exhibits and newly tested techniques. Throughout the book, welfare and wellness are used interchangeably, but we have attempted to delineate the differences between the two terms. Our approach in Jacksonville is aspirational, as we seek to provide elephants with a lively, ever-changing, naturalistic experience where autonomy and opportunity is evident each and every day of their lives. We are committed to spending as much money as it requires to reach this goal. At the same time, we will make every effort to advocate for elephants in the wild and keep our visitors informed about their plight and their progress.

As zoos morph from hard to soft architecture and bloom as verdant gardens that provide the context for wildlife, they also become oases for the surrounding community of humanity. Our most persistent critics want to close zoos even as they are successfully delivering the naturalistic habitats that encourage animals to thrive. At their best, zoos deliver psychological well-being for visitors and animals alike. Naturalistic zoos are urgently needed in urban centers where citizens have been isolated and estranged from the natural world (Maple & Morris, 2018). The best zoos are centers of learning and inspiration and islands of respite and revitalization. Zoos, when they meet the highest standards and best practices, should receive our full support. We must continue to advocate these standards for all zoos and turn away from the anachronistic, hard architecture that failed to meet the needs of elephants and other species in the zoo.

Architects, curators, directors, educators, trainers, and psychologists took part in this historic meeting in Jacksonville, Florida. Two of the papers are concerned with "otherkind" as we characterized the contributions that focused on giraffes and reptiles, respectively. These papers are included as a segue into the future of our scholarship. We intend to offer additional workshops to test the wellness construct and examine its utility for other taxa while providing a forum to describe and debate the research and design experience of others. It is especially important that we attract the best and brightest scholars to participate in our conversation. In the past, zoos have operated within an insular bubble of tired ideas. We can and must do better. As the foundation of the 2016 elephant workshop and future meetings is the idea that wellness is only achievable by the application of archival investigations of published research, systematic observations, and rigorous experimentation. We therefore present the findings of this workshop as a product of empirical zoo biologists and psychologists working in zoos committed to animal welfare in its highest form.

REFERENCES

Brambell, F. W. R. (1965). *Report of the technical committee to enquire into the welfare of animals kept under intensive livestock, husbandry systems* (The Brambell Report). London, United Kingdom: Her Majesty's Stationary Office.

Erwin, J., Maple, T. L., & Mitchell, G. (Eds.). (1979). *Captivity and behavior*. New York, NY: Van Nostrand Reinhold.

Fritz, J., & Howell, S. M. (1993). Psychological wellness for captive chimpanzees: An evaluative program. *Humane Innovations and Alternatives, 7*, 426–434.

Hancocks, D. (2009). What will new zoo environments look like? In D. L. Forthman, L. F. Kane, D. Hancocks, & P. F. Waldau (Eds.), *An elephant in the room: The science and well-being of elephants in captivity* (pp. 215-225). North Grafton, MA: Tufts University Press.Hanuliakova, J. (2015). Maslow's hierarchy of needs. In D.J Mellor, S. Hunt, & M. Gusset. *Caring for wildlife: The world zoo and aquarium animal welfare strategy.* (pp. 8). *WAZA Executive Office: Gland, Switzerland.*

Harlow, H. F. (1971). *Learning to love*. Chicago, IL: Aldine.

Harlow, H. F., Uehling, H., & Maslow, A. H. (1932). Comparative behavior of primates. I. Delayed reaction tests on primates from the lemur to the orang-outan. *Journal of Comparative Psychology, 13*(3), 313.

Hediger, H. (1955). *Studies of the Psychology and Behaviour of Animals in Zoos and Circuses*. London, United Kingdom: Butterworths Scientific Publications.

Hediger, H. (1970). *Man and animal in the zoo.* New York, NY: Delacorte Press.

Jensen, G. D. (1963). Preference for bar pressing over "freeloading" as a function of number of rewarded presses. *Journal of Experimental Psychology, 65*(5), 451–454.

Kane, L. F., Forthman, D. L., & Hancocks, D. (2009). Optimal conditions for captive elephants. In D. L. Forthman, L. F. Kane, D. Hancocks, & P. F. Waldau (Eds.), *An elephant in the room: The science and well-being of elephants in captivity* (pp. 346-379). North Grafton, MA: Tufts University Press.

Maple, T. L. (1996). The art and science of enrichment. In G. M. Burghardt, J. T. Bielitski, J. R. Boyce, & D. O. Schaefer (Eds.), *The well-being of animals in zoo and aquarium-sponsored research* (pp.79-84). Greenbelt, MD: Scientists Center for Animal Welfare.

Maple, T. L. (2016). *Professor in the zoo.* Fernandina Beach, FL: Red Leaf Press.

Maple, T. L., & Bloomsmith, M.A. (2018). Introduction: The science and practice of optimal animal welfare *Behavioural Processes,* 156, 1.

Maple, T. L., Bloomsmith, M. A., & Martin, A. L. (2009). Primates and pachyderms: a primate model of zoo elephant welfare. In D. L. Forthman, L. F. Kane, D. Hancocks, & P. F. Waldau (Eds.), *An elephant in the room: The science and well-being of elephants in captivity* (pp. 129-153). North Grafton, MA: Tufts University Press.

Maple, T. L., & Bocian, D. (2013). Commentary: Wellness as welfare. *Zoo Biology, 32*(4), 363–365.

Maple, T. L., & Morris, M. C. (2018). Naturalistic and natural settings and psychological well-being. In A. S. Devlin (Ed.), *The environment and human behavior: The effects of built and natural settings on wellbeing* (pp.253-279). Cambridge, MA: Academic Press.

Maple, T. L., & Perdue, B. M. (2013). *Zoo animal welfare*. New York, NY: Springer.

Maple, T. L., & Segura, V. D. (2015). Advancing behavior analysis in zoos and aquariums. *The Behavior Analyst, 38*(1), 77–91.

Maple, T. L., & Segura, V. (2018). Wildlife wellness: The next ethical frontier for zoos and aquariums. In B. Minteer, J. Maienschein, J.P. Collins. (Eds.). *The ark and beyond: The evolution of zoo and aquarium conservation* (pp. 226–237). University of Chicago Press.

Markowitz, H. (1982). *Behavioral enrichment in the zoo*. New York, NY: Van Nostrand Reinhold.

Markowitz, H. (2011). *Enriching animal lives*. San Francisco, CA: Mauka Press.

Maslow, A. H. (1962). *Toward a psychology of being*. New York, NY: Van Nostrand.

Mellor, D. J., Hunt, S., & Gusset, M. (2015). Caring for wildlife: The world zoo and aquarium animal welfare strategy. *WAZA Executive Office: Gland, Switzerland.*

Norton, B. G., Hutchins, M., Stevens, E., & Maple, T. L. (Eds.). (1995). *Ethics on the ark: Zoos, animal welfare, and wildlife conservation.* Washington, DC: Smithsonian.

Novak, M., & Suomi, S. J. (1998). Psychological well-being of primates in captivity. *American Psychologist, 43,* 765–773.

Novak, M. A., & Petto, A. J. (1991). *Through the looking glass: Issues of psychological well-being in captive primates.* Washington, DC: American Psychological Association.

Segal, E. F. (Ed.). (1989). *Housing, care, and psychological wellbeing of captive and laboratory primates.* Park Ridge, NJ: Noyes.

Tennant, K. S., Morris, M. C., Segura, V. D., Denninger-Snyder, K., Bocian, D., Lee, G. H., & Maple, T. L. (2018). Achieving optimal welfare for the Nile hippopotamus (*H. amphibius*) in North American zoos and aquariums [Special issue]. In T. L. Maple, & M. A. Bloomsmith (Eds.), *Behavioural Processes* (pp.51-57).

Yerkes, R. M. (1925). *Almost human.* New York, NY:

Yerkes, R. M. (1943). *Chimpanzees: A laboratory colony.* New Haven: Yale University.

Chapter 2

ELEPHANT WELLNESS

The Care of Asian Elephants (*Elephas maximus*) in Dublin Zoo as a Model for the Future

S. K. McGuinness, FRGS

INTRODUCTION

E lephants have colored human culture for millennia, from Hannibal's armies (probably *Loxodonta africana pharaoensis*) to Hindu worship of the elephant god Ganesha and their modern use as work animals in Asian timber camps. Thus, the relationship between today's zoo elephants and the public is perhaps accepted most easily in a highly scrutinized zoo world. The visiting public unquestionably expects elephants to be represented in a modern zoo, yet the intelligence and social complexity of these largest land animals cannot be understated. With growing public perception of the potential for compromised welfare in zoos and the questions whether some species should remain in captivity at all, a new direction for the zoo elephant status quo is long overdue. Using the example of Dublin Zoo's elephant wellness program and its quantifiable success, the following paper outlines a new direction for the status and conditions of elephants in a zoo environment that is centered on an approach to holistic care that transcends physiological metrics. This should not be considered a justification for elephants to exist exclusively in zoos. As Maple, McManamon, and Stevens (1995) observed, "Managers of captive animals should never fool themselves with the belief that they can replicate nature in a captive setting." On the contrary, my goal in this paper is to outline the way in which the foreseeable future of elephants in zoos can be maintained and optimized. To achieve this, however, requires a revised statement of best practices to represent a point of no return in animal wellness.

THE LEGACY OF ELEPHANT CARE IN ZOOS

In 1974, Robert Sommer first reframed the captive management of wild animals. By coining the phrase "hard zoos," he described the physically immobile housing for zoo animals that neither reflected their natural habitat nor provided them an opportunity to exhibit natural behavior. For elephants in these conditions, naturalistic activity budgets (particularly those of feeding and locomotion) were achieved rarely. Further, with the necessity to control elephants in conditions of full contact (FC) using an ankus to dominate the herd with force or fear of force, complicated (although largely understudied) social structures were modified that compromised cultural exchange within herds. In addition, up to 40% of zoo elephants suffered foot infections attributable to prolonged immobility on hard surfaces in holding areas that were often small. Not only did this cause pressure sores, but exposure to feces and urine caused serious infections and irregular growth known as "hoof cancer" (Klos & Lang, 1982). More recent assessments have found that a large proportion (31%) of zoo elephants still experience foot and joint problems linked directly to standing on hard housing surfaces and limited movement for extended periods (Mellen, 2013). This study also correlated these conditions with anecdotal claims of shortcomings in management, poor care, and improper hygiene.

In addition to the obvious physical effects of historical zoo elephant care, psychological distress also has been documented well. For example, in a recent review, almost half of the UK herd of zoo elephants was found to perform some form of stereotypical behavior for more than 5% of a 24-hour period (Harris, Harris, & Sherwin, 2008), and this correlated rarely with historical conditions. Thus, current conditions of some zoo elephants are perpetuating negative stereotypies beyond the naïve care conditions of the past.

PROTECTED CONTACT

The scenario above remained the *modus operandi* of zoo elephant systems for several centuries. Recently, some assessments have normalized FC in elephant care (Cohn, 2006), advocating the walking of elephants on a daily basis in place of the provision of larger zoo habitats. However, with the advent of more advanced psychoanalytical approaches to animal behavior and motivations, and with greater realization of the welfare constraints of raising elephants in a zoo environment, much has changed since. Not only have zoo elephant habitats become much larger and mimic wild habitat more functionally than ever before ("fewer species living large," as Maple (2012) stated succinctly), but more appropriate training techniques have been developed that reduce the necessity to dominate herd dynamics.

Of significant note in the evolution of zoo elephant handling systems was the development of protected contact (PC) training, which emerged in response to concerns for keeper safety and animal well-being. Through this system, keepers are no longer present within the herd, and the elephants' participation in husbandry activities is voluntary, enhanced by the use of exclusively positive reinforcement. Recent studies have quantified a significant increase in latency of choice (deemed to indicate volition) by elephants kept under these protected contact conditions (Wilson, Perdue, Bloomsmith, & Maple, 2015). As choice is now recognized as a prominent metric of animal well-being (or "positive affective states") in zoo environments (Whitham & Wielebnowski, 2013), this development is a significant step. PC also has led to a general increase in elephants' tolerance for unusual circumstances and events (Desmond & Laule, 1993), a trait that represents part of a suite of species-typical behaviors and undoubtedly enhances the behavioral resilience of popular zoo animals.

Although seemingly beneficial to keeping elephants in zoos, some recent studies have continued to recommend FC and negative reinforcement for elephants, insisting that this may be the only means of providing adequate exercise in confined environments. Some analysts have warned that this apparent siege mentality, one that likens PC to a "robotic, remote relationship" that erodes existing animal-keeper relationships, serves to slow the evolution of elephant welfare in zoos as a whole (Desmond & Laule, 1993). A more recent study has shown higher levels of elephant-keeper interaction under PC versus FC training (Wilson et al., 2015). One valid limitation to a PC system of improved animal welfare is the poorly addressed complex ecology in captivity. For example, studies of night movement have shown that nocturnal behavior should not be extrapolated from daytime activity budgets (Posta, Huber, & Moore III, 2013). Similarly, younger members of the herd represent the social structure of zoo elephants increasingly, in contrast to that in wild populations. A recent assessment of UK zoo elephant populations highlighted this as a pressing concern in elephant welfare (Harris et al., 2008). Further, the same report indicated that most indoor spaces represent a "minute fraction" of wild elephants' night ranges (Harris et al., 2008). Further, this study found no strong relationship in the expression of stereotypes and fecal cortisol metabolite (FCM), a reliable indicator of stress in zoo mammals, between handling methods (PC versus FC). Added to this, given the reality of working within the resource confines of zoos, a more holistic approach to elephant care is required, which complements the strides taken by adopting PC to fill the remaining welfare gap.

ELEPHANT WELLNESS

The operations of several zoos internationally appear to represent this phase shift by providing an extension to the often simplistic concept of animal "welfare" based on purely physiological metrics. Thus, so-called animal wellness has emerged and is framed as a holistic blend of biology and psychology that facilitates expression of wild elephants' group dynamics by linking PC, larger, and more naturalistic habitats and modified keeping approaches that are mindful of the bidirectional relationships between keepers and animals. Further, in conjunction with appropriate habitat design and planting, this philosophy is intended to provide a calming atmosphere for visitors that showcases healthy animal group dynamics (Maple, 2012).

As Maple, Bloomsmith, and Martin (2009) observed, this movement toward wellness in pachyderms followed closely the improved care of nonhuman primates in zoo environments because of concerns for their welfare, including the constructive dialogue (and criticism) involved in such a process (Maple & Perdue, 2013). As some of the movement's leading exponents, the authors went further by claiming that until quite recently, elephants have been left behind in the shift from the hard zoos of the past to the softer, immersive architecture of the present. Regardless of whether the past practices were perceived as a necessary evil of handling large and powerful animals, the shift to activity-based designs for elephants has begun in earnest (Coe, 1985; Maple & Perdue, 2013).

A recent assessment of zoo elephant welfare conducted by the RSPCA (Clubb & Mason, 2002 found that several conditions are likely to lead to various metrics of poor elephant welfare. These include, but are not limited to, restricted space and opportunities for exercise; cold, wet climates; hard, wet substrates; extended periods of confinement; hard and wet substrates; inappropriate diets; lack of opportunities to perform natural behaviors; small, unrelated, unstable social groups; early weaning, breaking, and exposure to aversive stimuli during training (see Table 1). Clearly, improvements are required. Although empirical data on the causal factors of such welfare limitations are lacking, this chapter contends that an elephant wellness approach addresses these drivers largely. Further, it is becoming increasingly clear that zoo populations cannot be sustained by captive breeding alone under existing circumstances (Schulte, 2000). In fact, until quite recently, most (57%) North

American zoos housed fewer than three female Asian elephants (Schulte, 2000). Thus, a breeding population that includes the ability to establish long-term social groupings successfully, and therefore increase breeding success, does not appear to have been a concern in recent zoo thinking.

Bearing this in mind, I present an overview of Dublin Zoo's Asian elephant program and highlight the advantages of adopting a "wellness" paradigm, for the benefit of elephants (and, by extension, a broader suite of species) kept in zoo environments.

Table 1

Recommendations for Continued Maintenance of Elephants in Zoos and How Dublin Zoo's Adoption of an Elephant Wellness Approach Has Addressed These Concerns

RECOMMENDATION	DUBLIN ZOO
Leaving males with their mother until the natural age of dispersal in the wild (10–15 years of age)	One male born and moved within a year, based on preexisting agreement. Current males (all < three yrs.) still with mothers. Continuity desired, if practicable
Leaving young females with their mothers for life	Current Dublin Zoo practice
Not housing any animal singly, especially females	Current Dublin Zoo practice
Not separating females from the herd for parturition, particularly when old, experienced females are present	Current Dublin Zoo practice
Adding enrichments, such as foraging devices, pools, rubbing/scratching posts, mud wallows, to indoor as well as outdoor enclosures	Current Dublin Zoo practice

RECOMMENDATION	DUBLIN ZOO
Not chaining, except for routine maintenance (e.g., bathing, foot care)	No chaining whatsoever: bracelet training conducted to facilitate emergency medical intervention, although unnecessary to date
Using rubber flooring in indoor enclosures	Current Dublin Zoo practice
Not housing elephants indoors for more than a few hours per day unless the space available indoors per elephant meets minimum recommendations for outdoor enclosures	Continuous outdoor access provided all day, barring brief periods of habitat servicing. Night access provided to outdoor habitat for 95% of the year (adverse weather conditions are only exception).
Revising diets to meet existing recommendations given by de Regt, Nijboer, & Bleijenberg (1998)	Current Dublin Zoo practice
Discontinuing all forms of breaking	No breaking in Dublin Zoo
Greater transparency and record keeping regarding the use of aversive stimuli during breaking and training, so that use can be monitored	No breaking in Dublin Zoo

DUBLIN ZOO: PAST

Established in 1831, Dublin Zoo remains one of the oldest in the world, following closely in the footsteps of Vienna (1752), Paris (1793), and London (1828). Further, it was among the first to offer access to the public from its inception and, thus, represents an important educational and cultural facility for Dublin and its visitors. As long ago as 1835, Dublin Zoo provided short-term care for (mainly Asian) elephants and received its first resident elephant in 1836. Through various phases of financial fortune, Dublin Zoo has housed elephants, usually as solitary individuals, all handled under FC conditions. This included the traditional "enclosure" approach with little acknowledgement of wild habitat or enrichment to provide opportunities to express species-typical behaviors (Figure 1). As of 2004, Dublin Zoo housed two wild-born Asian elephants in this system: Judy (born c. 1957) and Kirsty (born c. 1967).

FIGURE 1: THE ELEPHANT ENCLOSURE AT DUBLIN ZOO IN THE MID-1980S, DISPLAYING CONCRETE SURFACES, LIMITED HOUSING, AND LITTLE OPPORTUNITY TO ESTABLISH COMPLEX SOCIAL STRUCTURES. THE FIGURE SHOWS TWO AFRICAN JUVENILES, JUDY AND DEBBIE.

With increasing concern for keepers' safety (who were required to inspect the feet of our aging elephants regularly under FC) and to optimize the welfare of our elephants in a changing world, the decision was made to transition to PC under the guidance of Alan Roocroft, former Elephant Manager and Elephant Care Specialist at San Diego Zoo. However positive the strides made with our two elephants, current management expressed serious concerns about the inadequate facilities and lack of adherence to new European Association of Zoos and Aquaria

(EAZA) and the European Endangered Species Program (EEP) requirements to house retired elephants separately from breeding herds. Thus, a new elephant care philosophy, built from the ground up, was required.

STARTING AGAIN

Management realized quickly that Dublin Zoo possessed unique potential to revolutionize its elephant program, afforded by the releasing of additional space and funding and the requirement to relocate aging elephants. This also coincided with the availability of a breeding herd of Asian elephants residing in Rotterdam Zoo. In 2005, Judy and Kristy were relocated to Neunkircher Zoo, Germany, which allowed demolition of the existing infrastructure and construction of a new cowhouse, intermediate kraal, and outdoor habitat, all centered on a PC framework. The fundamental intention was to house a breeding group of elephants under a "wellness" philosophy. In 2006, the first members of our new group, two adult females (both pregnant upon arrival) and a three-year-old calf, arrived from Rotterdam Zoo. The matriarch of this family unit, Bernhardine, gave birth to a calf soon thereafter, followed by the birth of a second calf to Yasmin the following year.

The desire to maintain a self-sustaining, multigenerational breeding group resulted in the inclusion of a bull house and separate outdoor habitat in our redevelopment planning. This preempted new EAZA requirements stipulating that zoos holding breeding herds must provide facilities for a bull. Upali, the bull introduced, was chosen because his calm temperament was intended to improve breeding success and group cohesion, and the source institution, Chester Zoo, shared our breeding and care philosophy. Specifically, Upali's experience interacting with calves and breeding in their presence was a highly desirable trait. In 2012, Upali was delivered to his separate house, which was primed with dung from the cows and calves. Similarly, dung from Upali was scattered in the cows' outdoor habitat. Following a 24-hour period of observation showing normal characteristic behavior, Upali was introduced to the matriarchal herd in the outside habitat and, following 10 minutes of protective behavior from the matriarch, was subsequently accepted into the herd. Since his introduction, Upali has sired six calves successfully, and 11 healthy elephants now occupy our new elephant habitat (the firstborn male calf, Budi, was transferred to Denver Zoo at four years of age). Upali's interaction with both calves and cows remains close and extremely positive.

As Maple et al. (2009) indicated, a key method in achieving greater elephant wellness is ". . . to envision greater opportunities for wildness" in habitat design. Therefore, the redesign of our elephant habitat and housing was planned carefully to encourage a wide range of natural behaviors and facilitate world-class animal care with a wellness philosophy.

First, a deep sand substrate throughout the outdoor habitat allows much greater muscle activity within a limited area, particularly on the ascending/descending slopes, which are extensive throughout. Further, this sand is used regularly to display more naturalistic behaviors, including throwing, digging, and play. The animals also exercise a wider muscle set in lying on wet sand and standing again regularly. Second, two large pools have been provided within the wider habitat, both deep enough to allow free swimming and with divided access to prevent any one elephant from monopolizing the resource. All elephants use these pools extensively on a daily basis, and they have filtration systems that allow for minimal keeper intervention. A dry riverbed and mud bath area that are flooded periodically complement these and provide additional behavioral variation. Third, hanging feed nets positioned at key locations throughout the outdoor habitat provide varied and consistent resource movement during the elephants' time outdoors. Keepers control lowering these feed nets remotely, again reducing labor requirements and static feeding to a target of 15 minutes standing time. Fourth, the inclusion of a variety of natural features, including tree stumps and rocks that are moved and changed consistently, provides a varied naturalistic feel. Finally, this outdoor habitat is home not only to Asian elephants but is shared by blackbuck (*Antilope cervicapra*) that have night quarters separated by a soft boundary as well as free-roaming Indian peafowl (*Pavo cristatus*).

FIGURE 2: CURRENT ELEPHANT HABITAT AT DUBLIN ZOO, EXHIBITING
THE VARIOUS ELEMENTS OF OUR WELLNESS PHILOSOPHY.

As in the outdoor habitat, the indoor housing includes deep sand (180 centimeters) and hanging feed nets. In addition to the benefits to muscle and foot health provided by a soft substrate, this also offers opportunities to recline at night, increasing the elephants' durations of sleep. Further, elephants can interact as a group at night and often are given access to their outdoor habitat if weather permits. In addition, during cow oestrus, the bull Upali is provided with night access regularly. Sand is turned and piled daily, creating slopes that appear to be easier for elephants to sleep against. Again, tree stumps and rocks are provided, while keepers bury large boughs of browse deep each day to provide night forage. Feed holes built into the walls of cow and bull houses alike provide further naturalistic feeding, while a randomized time-release mechanism on these reduces the predictability in forage provision during the night.

A set of CCTV cameras, including infrared monitors, records night activity in both the indoor and outdoor areas (Figure 3). This has allowed breeding and calving to be monitored remotely without the need for keeper intervention and has provided data for multiple behavioral studies. Further, connection to a secure data network allows keepers to monitor the elephants on their mobile phones. As a fundamental component of our elephant wellness philosophy, a PC training wall runs along one side of a set of three training stalls adjacent to the indoor habitat. These have a rubber floor (the only nonsand substrate) and allow access for essential PC husbandry using a series of foot and ear ports. Finally, the indoor habitat has been fitted recently with a calf training crèche (CTC) to allow PC-based healthcare checks, particularly in light of the risk posed by elephant endotheliotropic herpes virus (EEHV). Designed with herd cohesion in mind, this allows tailored calf training to be conducted while the calf is separate, but adjacent to, its mother (Figure 4a).

The amount of space provided for our animals was considered carefully. A number of other facilities have praised designs that provide enough space to allow expression of a minimum allowable suite of behaviors (Jones & McGreevy, 2007). However, the goal of our redesign was not simply to achieve the minimum standards that ensure physical survival and health, but to surpass them and allow the display of large group behaviors (often disregarded in other studies). Indeed, a recent study has shown the advanced spatiotemporal and social abilities of elephants and their strong links to wide wild ranges (Hart, Hart, & Pinter-Wollman, 2008). Conversely, the mere amount of space might not be indicative of improved welfare (Maple et al, 2009). The quality of this space, the amount of stimulation afforded to the animals, and the frequency with which these positive design elements are used is of much greater importance (Burghardt, 1996; Meehan, Mench, Carlstead, & Hogan, 2016).

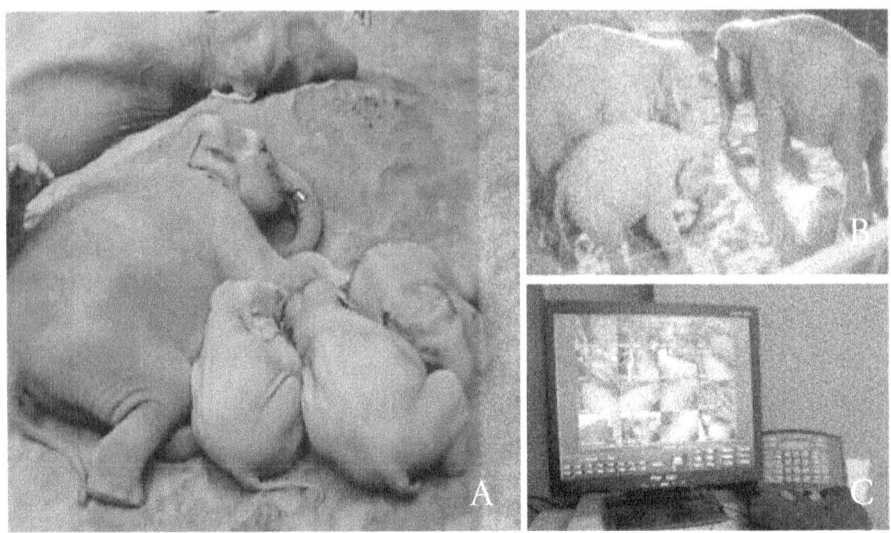

FIGURE 3 (A–C): DUBLIN ZOO ADULT FEMALE ASHA SLEEPS WITH THREE
CALVES, SAMIYA, KAVI, AND ASHOKA ON INTERIOR SAND PILE, 2015 (A). ASHA'S
NIGHTTIME BIRTH IN 2007 WITHOUT KEEPER INTERVENTION (B) SHOWED
SPECIES-TYPICAL HERD DYNAMICS. EXTENSIVE CCTV (C) ALLOWS CAREFUL
MONITORING OF BOTH PROCESSES WITHOUT KEEPER INTERACTION.

Although, as Maple et al. (2009) stated, "Nature is the superior mode
for zoo designers," zoos must remain financially viable through visitation revenue. In our wellness philosophy, the redesign by Jones and
Jones Architects Ltd. was informed by visitor experience as well as animal comfort, with "landscape immersion" that provides a sense of exploration and discovery. Much research in learning has revealed that this
form of exposure is a much more effective educational tool than traditional exposition. This facilitates the development of respect for the
habitat as well as the animals therein, through a holistic learning experience (Campos, 2015). In addition, as noise has been long-understood and
recently proven (Quadros, Goulart, Passos, Vecci, & Young, 2014) to affect zoo animal behavior, a design that reduces public noise was essential
for the new elephant habitat. Thus, limiting public viewing to strategically located cross-views and planting extensive bamboo barriers helped

both reduce daytime noise and provide extensive areas of animal privacy under the animals' own volition.

THE BENEFITS OF A WELLNESS PHILOSOPHY: FEEDING

In the wild, African elephants have been found to spend up to 75% of their time foraging (Posta et al., 2013), while Asian elephants may spend over 90% of their time feeding (McKay, 1973). In zoo environments, the provision of an unpredictable feeding regime has been related directly to reduced obesity (Morfeld, 2013), which also is related directly to the number of times food is made available per day. The continuous availability of forage within the elephant habitat in Dublin Zoo and the largely random nature of its provision have resulted in a highly mobile herd that feeds for the majority of the day and night. Feed boxes built into the walls of both cow and bull houses provide further opportunity for randomized night feeding. Further, reaching and stretching for high browse throughout their habitat has resulted in substantially improved body condition and muscle strength in the neck and shoulders, a benefit not usually provided by more traditional feeding enrichment.

MOVEMENT

With respect to movement, ranging behavior in the wild has been found to be related directly to resource availability, including water and forage (Williams, 2009). Thus, based on physiology alone, provision of sufficient amounts of these resources in a zoo environment, together with many opportunities to exercise, may not necessitate large ranges in zoos. Because of the feeding regime outlined above, and the availability of a large number of opportunities to express other natural behaviors, the movement of our zoo animals is on a par with wild herds and is conducted on species-typical substrate. A recent sample of US zoo elephants (n=240) found that 74% were overweight/obese, with higher average figures observed for female Asian elephants (Morfeld, 2013). Similarly, UK zoo elephants spent less time feeding than did wild elephants but also were overweight, indicating that diet and/or lack of exercise probably were contributing factors (Harris et al., 2008). However, the recent Dublin Zoo locomotion study found that our female group travels an average of 9.32 kilometers/day and our bull travels an average of 14.4 kilometers/day (Brady, 2015). This indicates that levels of exercise in our resident animals are comparable to those among wild groups (c), a finding similar to that in the Melbourne Zoo PC herd (Rowell, 2014). Further, the weights of our elephants are consistent with those of wild elephants of comparable ages. Given that Clubb and Mason (2002) have related obesity and stress directly to impaired survivorship, our Asian elephants can be expected to live longer, although direct comparison cannot be made at this early stage of our wellness program.

SOCIALITY

The species-typical behavior of Asian elephants includes complex and extensive social structures, most of which are less understood than those of primates. In contrast to African elephants, however, recent observations have found that Asian elephant groups have less connectivity at the population level and are affected less by annual ecological change (de Silva & Wittemyer, 2012). However, this study also predicted that Asian elephant social structure would fragment more quickly with the loss of group members. Thus, maintaining smaller, close-knit groups is of the utmost importance in promoting healthy species-typical Asian elephant groupings in a zoo environment. Since the revolution in our elephant care philosophy, we have observed marked improvements in sociality. A sleep study in 2012–13 found that our elephants sleep lying down on average three and a half hours per night (Walsh, 2017). Calves sleep regularly for substantially longer periods, and a large proportion of this sleep is in contact (Figure 3a). This is interpreted to be a direct result of the use of sand and the facilitation of group activity during the night. Similarly, although some of our elephants originate from an FC environment and do occasionally display stereotypies, an increase in space and interaction with a diverse conspecific social profile (particularly at night) has led to reductions in these displays, as observed in similar case studies (Greco, 2013). Further, our elephants' rate of vocalization is higher than in other zoos and shows a repertoire unrecorded previously (M. Artelt, personal communication, September 1, 2008).

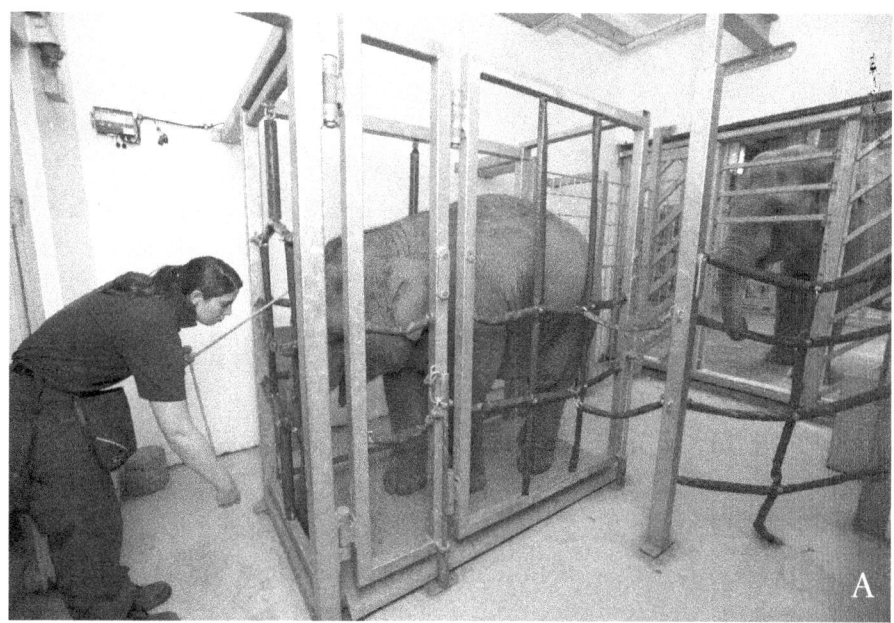

FIGURE 4 A AND B: (A) CALF TRAINING CRÈCHE (CTC) IN DUBLIN ZOO'S ELEPHANT COWHOUSE, SHOWING KAVI RESPONDING TO POSITIVE REINFORCEMENT, WITH OPEN ACCESS TO THE INDOOR HABITAT. (B) KAVI RECEIVES TRAINING IN ADMINISTRATION OF RECTAL FLUIDS FOR TREATMENT OF EEHV. NOTE ENTHUSIASTIC VOLITION TO PARTICIPATE IN BOTH. SEE SUPPLEMENTAL VIDEO MATERIAL AT THE FOLLOWING PERMALINK: HTTP://BIT.LY/2BMCCQX.

HEALTHCARE

A fundamental component of elephant wellness is well-managed health-care. As an increasingly recognized metric of good animal welfare in zoos (Whitham & Wielebnowski, 2013), especially with respect to elephants who are at risk of contracting potentially-fatal infection loads (e.g., EEHV and calves), epidemiology is of paramount concern in Dublin Zoo's elephant wellness program. The majority of PC training is conducted now as a way to facilitate physiological monitoring and future interventions if required. The adoption and improvement of PC training techniques allow for continual health checks for symptoms of EEHV and response to infection, if required. Continuous improvement in PC access to animals (particularly with highly vulnerable calves) and exceeding international best practices facilitate this. The absence of infection in Dublin Zoo demonstrates improved welfare by indicating reduced stress and associated strengthening of immune systems, while the training of preventative interventions may, if required, save the lives of PC-trained elephants (Figure 4b). PC and the design features of our elephant habitat also have provided substantial foot-care benefits. As Asian elephants require more nail care than do their African counterparts, the benefits of regular PC foot-care interventions have been highly apparent in Dublin Zoo, as in other institutions (Roocroft & Oosterhuis, 2001). In the 10 years since we established our wellness program, not a single foot issue has been detected, a marked deviation from the norm (Mellen, 2013). Having said this, overall longevity remains significantly higher in wild elephants. In a comparison of wild-caught Asian elephants working in timber camps and zoo-based Asian elephants, Clubb and Mason (2002) found that the median adult life span in zoos was circa 18.9 years, while Asian elephants working for timber companies lived an average of 41.7 years. In addition, infant mortality was higher in zoo elephants. That said, given the youth of PC programs in zoos around the world and the very recent advent of wellness approaches, longevity in our subset of zoo elephants cannot yet be compared to those in the wild, although it is expected to exceed current norms.

BREEDING

The majority of elephant pregnancies in North American zoos that house elephants are achieved via artificial insemination (Maple et al., 2009). In Dublin Zoo, Upali and the females copulate regularly, resulting in the birth of six calves. These events have been observed by calves and other adult females, building knowledge and strengthening bonds, and our animal care team also has monitored them remotely through CCTV. The use of sand, which cushions impact, absorbs associated fluids, and provides a grip for calves attempting to stand, has improved the delivery of these new calves substantially. Similar to observations in Chester Zoo and Hagenbeck's Tierpark, Hamburg, newborn calves in Dublin Zoo stand within six minutes on sand by comparison to approximately 30 minutes on concrete surfaces. This process reduces the stress levels of the mother and the wider herd. Anecdotally, this has had important ramifications for the independence and strength of our calves as they grow (G. Creighton, personal communication, August 8, 2016). The healthy weight of our cows may also contribute to our 100% birthing success, as suggested by Olson (2004). Furthermore, an unexpected benefit of our breeding program has been the sex division of our calves. With three males and three females, males focus their increasing levels of testosterone on physical play, leaving females time and space to develop the complex social relationships they will need in later life. This serendipitous outcome also reduces anxiety levels in calves and, thus, exposure to stress-related diseases such as EEHV and arthritis (Clubb & Mason, 2002). With respect to fecundity, Dublin Zoo female elephants now match wild Asian elephants, who give birth to roughly one calf every six years for almost their entire adult lives (Mason & Veasey, 2010). This is likely related to reduced stress levels in Dublin Zoo, although other factors also may have an effect (Mason & Veasey, 2010). Dublin Zoo now cares for the oldest elephant calf in the world born into an exclusively PC system, nine-year-old Asha, who has recently given birth to her own calf.

RESEARCH

The process of reinventing our elephant care program also has provided the tools to undertake vital behavioral, dietary, and healthcare studies (Whilde & Marples, 2012). As several zoo researchers have noted, the level of zoo-based research requires considerable improvement to yield quantifiable standardization of care (Barber, 2009; Clubb & Mason, 2002; Maple et al., 1995).

VISITOR EXPERIENCE AND EDUCATION

Finally, in addition to welfare improvements achieved through the adoption of an elephant wellness philosophy, this approach and the associated development of our Kaziranga Forest Trail has provided a powerful visitor experience, including various cross-views for independent animal discovery, sensory stimulation through phytogeographical planting (Butler, 2016), and wide potential for educational engagement through the display of naturalistic behavior (Stroud, 2007).

Overall, Clubb and Mason's 2002 report on the welfare of zoo elephants commissioned by the RSPCA raised several key concerns. Of their 11 immediate recommendations to improve elephant welfare in zoos, listed in the table below, Dublin Zoo has met all of these using a wellness framework (Table 1). However, we recognize that this is not an end point and are committed to continuous improvement.

As is apparent, the wellness philosophy applied with the elephant herd of Dublin Zoo, which our world-class infrastructure and expertise support, is successful. Maintaining and improving upon these advances remains a priority.

THE FUTURE

Dublin Zoo's role in the future of elephant wellness is linked intrinsically to its successful development of PC and the maintenance of a habitat that allows the expression of naturalistic behavior. Given the success of breeding at the zoo and the highly social nature of our group (despite the inherent resource limitations of a city zoo), it will become necessary to split our herd soon. Since the birth of the three most recent calves, two subgroups have begun to emerge within the herd: two adult females with three female calves, and two adult females with three male calves. It is intended to maintain and strengthen this subdivision while retaining cow-calf dyads before an intended move once calves are established (circa two years). The receiving institution works closely with the staff of Dublin Zoo, ascribes fully to our PC philosophy, and has plans to include a breeding bull in their herd. Further, their newly built infrastructure has been modeled upon our own and exceeds ours in size. This presents an opportune moment to reassert the fact that the Dublin Zoo elephants should represent the future of zoo elephants as part of a founder population for a PC-trained global zoo herd. Dublin Zoo's elephant calves are now the oldest in the world to be reared entirely under PC from birth, and although speculation on the limits of their learning ability remains exactly that, these calves represent the potential seeds of a global elephant wellness program centered on the principles of PC and sociopsychological well-being. The level of future cultural exchange between PC calves and adults that have a FC legacy is less clear, however. Although Clubb and Mason (2002) maintained that "bringing elephants into zoos profoundly impairs their viability," the results of our program to date suggest strongly that a wellness approach to elephant care has the potential to change this outcome.

Maintaining fidelity to the values of wellness may become challenging as time progresses and institutions come under increasing pressure to conform to public opinion (Cohn, 2006) and policy guidelines (e.g., Defra, 2012). For example, combinations of PC and FC have emerged that involve a PC wall and restraint bracelets, together with subsequent keeper interaction (Desmond & Laule, 1993). We do not believe this mixed form of PC and FC is faithful to a wellness philosophy. Further, the within-zoo politics of PC and the challenges of negotiating what is sometimes a siege mentality have been noted (Desmond & Laule, 1993; Whitham & Wielebnowski, 2013). The future of elephant wellness depends on institutional and keeper-level understanding of its values and the availability of staff trained appropriately who have high job satisfaction and exhibit a culture of continual reflective improvement. As with any major shift in zoo philosophy, the availability of sufficient finances and facilities is vital, while the uncompromising standardization of PC and elephant wellness as a whole is imperative to achieve systemic changes in the welfare of all zoo elephants (A. Roocroft, personal communication, July 14, 2016).

As was apparent in preparing this chapter, the paucity of peer-reviewed publications in this field will continue to hinder the development of wellness beyond private consultancy and personal experience. Recent reviews of elephant welfare in zoos support this assertion (Barber, 2009; Clubb & Mason, 2002; Meehan et al., 2016). If genuine standardization of conditions and resilience to disease are sought, I recommend increasing specific empirical investigation of the benefits of a wellness philosophy; using identified key performance parameters, such as longevity, fecundity, infant mortality, activity budgets, and various physical health metrics; as well as providing uncompromising conditions to support this.

CONCLUSION

Alan Roocroft asserted that we should avoid the temptation to declare a destination in elephant wellness programs, but simply maintain forward momentum. Although I agree with this assertion, care is required to ensure this phrase does not become an ideological silver bullet for welfare concerns: it must not remain in the abstract realm of intellectual theory. Similarly, we should not deem this a "terministic screen" from the pragmatic concerns of large animal care. Rather, adopting elephant wellness protocols should set a precedent from which to depart that passes the point of no return on the poor practices of the past and ensures a sustainable zoo elephant population for the future. Dublin Zoo has achieved the metrics of elephant welfare (longevity, expression of species-typical behavior and physiology, and successful breeding). Moving forward, a philosophy of elephant wellness strengthens these advances and establishes a paradigm shift in care.

As Maple et al. (1995) stated, all zoo practitioners have the duty to endeavor to make good zoos the rule, not the exception, based on the pillars of conservation, education, science, and recreation. Providing a holistic approach to animal care which necessitates (or gives opportunities for) animals to adopt naturalistic behavior is fundamental in achieving this goal. Concurrently, this provision ensures more effective communication of the right conservation message—the inspiration that zoos can provide through firsthand observation of naturalistic habitats occupied by healthy animals.

ACKNOWLEDGMENTS

I would like to acknowledge the hard work of the entire Dublin Zoo team in continuing to ensure the success of our elephant wellness program, especially our elephant care team, Operations Manager Gerry Creighton, the discovery and learning department, and Assistant to the Director Paul Donoghue. In addition, the kind support of Rotterdam and Chester Zoos has been instrumental in establishing our own herd. Jones and Jones Architects Ltd. deserve great thanks for designing a habitat that allows our elephants to express wild behavior while providing a valuable educational resource. Finally, Alan Roocroft's continued support and advice in developing our PC training has been fundamental to our success.

REFERENCES

Barber, J. C. E. (2009). Unpacking the trunk: Using basic research approaches to identify and address captive elephant welfare concerns. In D. L. Forthman, L. F. Kane, D. Hancocks, & P. F. Waldau (Eds.), *An elephant in the room: The science and well-being of elephants in captivity* (pp. 111–128). North Grafton, MA: Tufts University Press.

Brady, A. (2015). *Locomotion in Asian elephants (Elephas maximus) at Dublin Zoo.* University College Dublin.

Burghardt, G. M. (1996). Environmental enrichment or controlled deprivation? In G. M. Burghardt, J. T. Bielitski, J. R. Boyce, & D. O. Schaefer (Eds.), *The well-being of animals in zoo and aquarium-sponsored research* (pp. 91–101). Greenbelt, MD: Scientists Center for Animal Welfare.

Butler, S. (2016). Plants in zoos. *The Plantsman, 15*(1), 39–43.

Campos, M. (2015). Zoos and aquariums. *Attractions Management, 3,* 64–69.

Clubb, R., & Mason, G. (2002). *A review of the welfare of zoo elephants in Europe* (A Report Commissioned by the RSPCA, University of Oxford, United Kingdom).

Coe, J. C. (1985). Design and perception: Making the zoo experience real. *Zoo Biology, 4*(2), 197–208. Retrieved from http://doi.org/10.1002/zoo.1430040211.

Cohn, J. P. (2006) Do elephants belong in zoos? *BioScience, 56*(9), 714–717. Retrieved from http://doi.org/10.1641/0006-3568(2006)56[714:DEBIZ]2.0.CO;2.

de Regt, A. C., Nijboer, J., & Bleijenberg, M. C. K. (1998). *Diet inventory European Asian Elephant (Elephas maximus)*. Rotterdam, Netherlands: EEKMA.

Defra. (2012). *Secretary of state's standards of modern zoo practice*. Bristol, United Kingdom.

De Silva, S., & Wittemyer, G. (2012). A comparison of social organization in Asian elephants and African savannah elephants. *International Journal of Primatology, 33*(5), 1125-1141.

Desmond, T., & Laule, G. (1993). The politics of protected contact. In *Proceedings of the American Association of Zoological Parks and Aquariums Central Annual Conference* (pp. 12–18). Omaha, NE: PUBLISHER GOES HERE.

Greco, B. J., Meehan, C. L., Hogan, J. N., Leighty, K. A., Mellen, J., & Mason, G. J. (2016). The days and nights of zoo elephants: Using epidemiology to better understand stereotypic behavior of African savannah elephants (*Loxodonta africana*) and Asian elephants (*Elephas maximus*) in North American zoos. *PLoS ONE*.

Harris, M., Harris, S., & Sherwin, C. (2008). *The welfare, housing, and husbandry of elephants in UK zoos*. Bristol, United Kingdom: Report to *DEFRA. University of Bristol*.

Hart, B. L., Hart, L. A., & Pinter-Wollman, N. (2008). Large brains and cognition: Where do elephants fit in? *Neuroscience and Biobehavioral Reviews, 32*(1), 86–98. Retrieved from http://doi.org/10.1016/j.neubiorev.2007.05.012.

Jones, B., & McGreevy, P. (2007). How much space does an elephant need? The impact of confinement on animal welfare. *Journal of Veterinary Behavior: Clinical Applications and Research, 2*(6), 185–187. Retrieved from http://doi.org/10.1016/j.jveb.2007.06.001.

Klos, H.-G., & Lang, E. M. (1982). *Handbook of zoo medicine*. New York, NY

Maple, T. L. (2012). A zoo where animals come first. *Observer, 25*(4).

Maple, T. L., Bloomsmith, M. A., & Martin, A. L. (2009). Primates and pachyderms: A primate model of elephant welfare. In D. L. Forthman, L. F. Kane, D. Hancocks, & P. F. Waldau (Eds.), *An elephant in the room: The science and well-being of elephants in captivity* (pp. 129–153). North Grafton, MA: Tufts University Press. Retrieved from https://www.scribd.com/doc/141932439/An-Elephant-in-the-Room.

Maple, T. L., McManamon, R., & Stevens, E. F. (1995). Animal care, maintenance, and welfare. In B. G. Norton, M. Hutchins, E. Stevens, & T. L. Maple (Eds.). *Ethics on the ark: Zoos, animal welfare, and wildlife conservation*. (219–234). Washington, DC: Smithsonian Institution Press.

Maple, T. L., & Perdue, B. M. (2013). *Zoo animal welfare* (1st ed.). Berlin, Germany: Springer-Verlag. Retrieved from http://doi.org/10.1007/978-3-642-35955-2.

Mason, G. J., & Veasey, J. S. (2010). What do population-level welfare indices suggest about the well-being of zoo elephants? *Zoo Biology, 29*(2), 256–273. Retrieved from http://doi.org/10.1002/zoo.20303.

McKay, G. M. (1973). Behavior and ecology of the Asiatic elephant in Southeastern Ceylon. *Smithsonian Contributions to Zoology, 125*, 113. Retrieved from http://doi.org/10.5479/si.00810282.125.

Meehan, C. L., Mench, J. A., Carlstead, K., & Hogan, J. N. (2016). Determining connections between the daily lives of zoo elephants and their welfare: An epidemiological approach. *PLOS ONE, 11*(7), e0158124. Retrieved from http://doi.org/10.1371/journal.pone.0158124.

Mellen, J. (2013). Foot and joint health. In A. Baker (Ed.), *Using science to understand zoo elephant welfare: Proceedings of the AZA National Conference).* Kansas City, KS:

Morfeld, K. (2013). Body condition. In A. Baker (Ed.), *Using science to understand zoo elephant welfare: Proceedings of the AZA National Conference* (pp. x–x). Kansas City, KS.

Olson, D. (2004). *Elephant husbandry resource guide.* Azle, TX: International Elephant Foundation. Retrieved from http://www.elephantconservation.org/iefImages/2015/06/CompleteHusbandryGuide1stEdition.pdf.

Posta, B., Huber, R., & Moore, D. E., III. (2013). The effects of housing on zoo elephant behavior: A quantitative case study of diurnal and seasonal variation. *International Journal of Comparative Psychology, 26*(1), 37–52. Retrieved from https://escholarship.org/uc/item/0z8722ss.

Quadros, S., Goulart, V. D. L., Passos, L., Vecci, M. A. M., & Young, R. J. (2014). Zoo visitor effect on mammal behavior: Does noise matter? *Applied Animal Behaviour Science, 156*, 78–84. Retrieved from http://doi.org/10.1016/j.applanim.2014.04.002.

Roocroft, A., & Oosterhuis, J. (2001). Foot care for captive elephants). In B. Csuti, E. L. Sargent, & U. S. Bechert (Eds.), *The elephant's foot* (pp.21-52). Iowa State University Press. Retrieved from http://doi.org/10.1002/9780470292150.ch5.

Rowell, Z. (2014). Locomotion in captive Asian elephants (Elephas maximus). *Journal of Zoo and Aquarium Research, 2*(4), 130–135.

Schulte, B. A. (2000). Social structure and helping behavior in captive elephants. *Zoo Biology, 19*, 447–459. Retrieved from http://doi.org/10.1002/1098-2361(2000)19.

Sommer, R. (1974). *Tight spaces*. Englewood Cliffs, NJ: Prentice Hall.

Stroud, P. (2007). Defining issues of space in zoos. *Journal of Veterinary Behavior: Clinical Applications and Research, 2*(6), 219–222. Retrieved from http://doi.org/10.1016/j.jveb.2007.10.003.

Walsh, B. (2017). Asian elephant (*Elephas maximus*) sleep study—Long-term quantitative research at Dublin Zoo. *Journal of Zoo and Aquarium Research, 5*(2). doi: 10.19227/jzar.v5i2.174

Whilde, J., & Marples, N. (2012). Effect of a birth on the behavior of a family group of Asian elephants (Elephas maximus) at Dublin Zoo. *Zoo Biology, 31*(4), 442–452. Retrieved from http://doi.org/10.1002/zoo.20408.

Whitham, J. C., & Wielebnowski, N. (2013). New directions for zoo animal welfare science. *Applied Animal Behavior Science, 147*(3), 247–260. Retrieved from http://doi.org/10.1016/j.applanim.2013.02.004.

Williams, A. C. (2009). Space use by Asian elephants (Elephas maximus) in Rajaji National Park, North West India: Implications for elephants held in captivity. In D. L. Forthman, L. F. Kane, D. Hancocks, & P. F. Waldau (Eds.), *An elephant in the room: The science and well-being of elephants in captivity* (pp.39-52). North Grafton, MA: Tufts University Press.

Wilson, M. L., Perdue, B. M., Bloomsmith, M. A., & Maple, T. L. (2015). Rates of reinforcement and measures of compliance in free and protected contact elephant management systems. *Zoo Biology, 34*(5), 431–437. Retrieved from http://doi.org/10.1002/zoo.21229.

Chapter 3

ELEPHANT LANDS—SUPPORTING THE NATURAL DYNAMICS OF AN ASIAN ELEPHANT HERD

Bob Lee, Sharon Glaeser, and the Oregon Zoo Elephant Care Staff

INTRODUCTION

Elephant Lands has been designed to support the natural dynamics of our Asian elephant herd. The new habitat exemplifies the Oregon Zoo's philosophy that all animals should have the opportunity to exert choice and self-determination. Using information gathered through more than 60 years of working with Asian elephants, we tried to see the world from an elephant's perspective, and we looked at the new habitat through their eyes as much as humanly possible. To state it simply, our goal was to construct a habitat that would meet the herd's biological, social, physiological, and psychological needs for generations to come. We were convinced that the elephants themselves and the way they live in their natural habitats would provide the best indicators for how we should shape the new design.

DESIGNING THROUGH THEIR EYES

Decades of research, hands-on experience, and an understanding of our elephants' individual needs informed and guided the design of Elephant Lands. Our animal-management goals for Elephant Lands were to encourage activity, promote a diverse range of natural behaviors, offer increased opportunities for choice and social interaction, and provide biologically meaningful choices and challenges. To promote a full range of natural behaviors, the habitat needed to include everything the herd would need to thrive—space to roam; choice and control in their daily lives; the ability to live in multigenerational matrilineal groups, which bulls can join or leave as they would in free-ranging populations (McKay, 1973); and the opportunity for young males to learn from older males.

The importance of providing zoo animals with appropriate challenges (Meehan & Mench, 2007) and the related variables of choice and control (e.g., Shepherdson, Carlstead, Mellen, & Seidensticker, 1993) through environmental enrichment (Shepherdson, Mellen, & Hutchins, 1999) has been convincingly documented by studies and publications on a wide variety of species including elephants (e.g., Shepherdson, 1999). This is further validated by the results of a recently published multi-institutional elephant welfare study which found that enrichment diversity was one of the factors correlated with less likelihood of hyperprolactinemia and more likelihood of normal reproductive cycling in African elephants (Brown et al., 2016); and feeding method diversity was correlated with increased movement in both African and Asian elephants (Greco et al., 2016; Holdgate et al., 2016). In this multi-institutional study, social and management factors were found to be important for multiple indicators of welfare, while habitat size was found to be less important than expected (Meehan, Mench, Carlstead, & Hogan, 2016); however, increasing the size of our habitat and creating a more flexible space remained important for supporting our animal management goals.

Social management of our elephants was a priority in the design of Elephant Lands. In the past 30 years, females (related and unrelated) at the Oregon Zoo were housed as a single group or as two separate groups comprising two to four individuals, with each group having a single dominant individual. Two to four adult bulls were present at any given time and were housed separately, except when placed with females for breeding or socialization. Historically, once young bulls were pushed out of their herds, many Western facilities housed them separately to protect them from the cows and other bulls. However, exposure to other bulls plays a role in the social and physiological development of males into adulthood (Evans & Harris, 2008; Rasmussen & Krishnamurthy, 2000). In a typical free-ranging elephant herd, male calves are exposed to adult males early in life, as dominant bulls come into the herd to breed cycling females (McKay, 1973). As a young male develops, he learns his place in the hierarchy (Evans & Harris, 2008) and learns how a breeding male interacts with a herd. As he approaches sexual maturity, he may roam farther from his mother, and the females begin forcing him to leave the security of their natal herd (Douglas-Hamilton, 1972; Moss & Poole, 1983; Sukumar, 1989), at which point he needs to form bonds with other males (Evans & Harris, 2008; Eisenberg, McKay, & Jainudeen, 1971). In free-ranging African elephants it was reported that male calves appeared to enjoy an adult bull's company, "hero worshiping" them (Christo, Wilkinson, & Sheldrick, 2009); and there is some suggestion that rearing bulls in the presence of an adult bull minimizes aggressive behaviors (personal communications). Our most recent male calf at the Oregon Zoo was given opportunities to safely interact with an adult bull starting at two and a half years of age—and, at Elephant Lands, we wanted male elephants to move through a variety of socially compatible situations with other bulls and cows, thus providing them a socially and psychologically stimulating environment throughout their entire lives.

To inform decisions as we moved through the design process, a team of animal-care staff, managers, architects, and builders worked together to develop our "nonnegotiables": the elements we believed to be critical to the success of the new habitat. These nonnegotiables informed our decision-making throughout the design process and became critical during the "value engineering" phase, when all of the ideas were reviewed to ensure they could be supported by the established budget. This helped ensure our vision for Elephant Lands could take shape and helped us "stay our course" when difficult choices needed to be made.

THE OREGON ZOO'S FIVE NONNEGOTIABLES

1. Self-determination and choices.

2. Support for the development of a multigenerational matrilineal herd, including giving males the ability to come into and out of the herd regularly.

3. Natural substrates throughout.

4. Opportunities to forage for 14–16 hours a day.

5. Allowing elephants to have access to the outside year-round (unless the temperature falls below 32°F).

DESIGN CRITERIA

Using the nonnegotiables, we developed our design criteria. In order to give the elephants choice, we looked at what motivates them naturally—searching out food and water sources, shelter, warmth, cooling, and engagement. We focused on their hierarchy of needs (Maslow, 1943; Mellor, Hunt, & Gusset, 2015), knowing this would encourage the animals to utilize the habitats without direct intervention by animal-care staff. We maintained focus on providing a physiologically, psychologically, and socially fulfilling environment with choice throughout the 24 hours in each day, including times when elephant staff are not working.

HABITAT/RESOURCE USE

The habitat's flexibility allows elephants to make decisions about where and how they spend their time and who they spend it with. Elephant Lands spans 6 acres with 4.5 acres of habitat. Its shape and topographical complexity encourage exploration and activity—the elephants cannot see the entire space from any one vantage point and get exercise simply by maneuvering through it (Figure 1). The perimeter walking distance is more than 1.3 miles, and the hilly terrain, various climbing features, and deep sand provide stimulation and physical challenges. Natural substrates, at least 4 feet deep throughout the facility, include a specific sand type (USGA top dressing) that cushions the elephants' feet and allows for drainage, natural dirt, and grass. A 160,000-gallon pool allows the entire herd to swim together if they choose (Figure 2). Other water features include a wading pool, drinkers, and a water cannon that can also be used to make a mud wallow (Figure 3). When elephants want to find food, water, or shelter, they have to actively look for it. To encourage this foraging behavior, we placed 25 feeding stations throughout the habitats, ranging from timed feeders that release food at programmable intervals to overhead boom feeders and large concrete herd feeders that are randomized daily (Figure 4). Inside Forest Hall, an activity tree features a timed feeder and mesh boxes with food items for elephants to explore; food items can also be hung from a hoist system and moved around the indoor space (Figure 4). An elephant may explore one area of the habitat and find no food only to return an hour later and find a small snack or even a "buffet."

FIGURE 1: ELEPHANT LANDS, (A) RENDERING FOR OVERALL SITE
PLAN (MARCH 2013), (B) ELEPHANT LANDS HABITAT.

FIGURE 2: 160,000-GALLON SWIMMING POOL.

FIGURE 3: WATER FEATURES INCLUDING A WADING POOL,
WATER CANNON, AND DRINKERS (NOT SHOWN).

FIGURE 4: FEEDING SYSTEMS (A AND B), AUTOMATIC FEEDERS (C AND D), BOOM FEEDER (E AND F), ACTIVITY TREE, AND FOOD OBJECTS HUNG FROM A HOIST SYSTEM, CONCRETE FEEDERS (NOT SHOWN).

Staff can also adjust heaters and water misters remotely. If elephants are staying under one shelter for a prolonged period of time, we can lower the temperature there, encouraging them to seek out warmth at another heater that is turned up in a different part of the habitat. "Deluge systems" at all of the shade structures and indoor elephant-care areas, including the nearly 12,000-square-foot Forest Hall herd room (Figure 5), rinse the sand and keep it clean—with the added benefit of being able to make it "rain," indoors or out (Figure 5). The elephants are free to seek out conditions they prefer, and the facility's complexity lets us provide a variety of opportunities from which they can choose.

All of this, it is important to note, can be accomplished without the direct intervention of the staff, so the elephants are no longer waiting by a door to be let in or fed. Instead, we manipulate resources in a fashion that mimics what can be found in some of their natural habitats. This environmental complexity helps evoke natural responses and behaviors in the elephants. It is up to them to explore and find what they need independent of the staff.

FIGURE 5: ENVIRONMENTAL CONTROL (A), FOREST HALL (B) SHADE STRUCTURES.

SOCIAL DYNAMICS

Knowing that we may be providing homes for new animals in the future—and that we want to provide more social opportunities for male elephants—we designed part of the building to function as an introduction space. A quarantine space (Figure 6) is negatively air-pressured, while the rest of the building is positively air-pressured. Within the quarantine space, we have six air changes per hour, and the air is exhausted more than 30 feet from any intakes, ensuring there is no cross-contamination. The quarantine area also includes an outside yard with an automatic feeder and an upright tree trunk from which to hang enrichment and food objects.

FIGURE 6: QUARANTINE (A), CORRAL (B) IN FOREST HALL
WITH A JUVENILE MALE AND ADULT MALE

A "howdy" system allows us to systematically bring new members into our herd and allows male elephants to enter and exit the herd with varying levels of physical interaction. The first level of the howdy system includes two barn stalls with a 10-foot space between them, which allows visual access between elephants without any physical contact. The second howdy level is a stall with a door system that allows keepers to control the extent of physical interaction. A mesh door on an extended overhead beam in the keeper space outside the stall can be attached to the solid concrete door when it is fully closed and pulled into the space as the concrete door is opened. The mesh is attached to the frame of the door in four rectangular panels, one on top of the other, which can be removed once we are confident in elephant interactions through the one-inch-by-one-inch mesh. The first step is to remove the top panel so elephants are able to touch up and over through the top opening. Next steps are to remove lower panels to allow them to touch lower on the body with greater reach. The final howdy level is to move the elephant into the "corral" in Forest Hall (Figure 6). The corral is a 1,600-square-foot space with vertical poles and doors on each end. With the doors closed, the elephants can have full contact and still be able to move away from one another if they choose. Once the elephants are ready, the two doors leading into the corral are opened, allowing for circle access and for the herd dynamics to then take over.

Allowing for the introduction and separation of individuals is critical to the successful management of the herd dynamics. By maintaining species-appropriate social groups, we can encourage a wide range of social behaviors. These behaviors, combined with the size and complexity of Elephant Lands, promote movement in our varied terrain and enhance individual, elephant-induced exercise.

The layout of these spaces allows elephants to choose how to socialize, giving bulls the opportunity to come in and out of the herd and for females to stay together as a group or spend time alone, choosing their social partners throughout the day.

ASSESSMENT OF ELEPHANT LANDS

We are in the analysis phase of a four-year study to scientifically assess the effectiveness of Elephant Lands. By collecting quantitative behavioral and physiological data on all of the elephants (Table 1), the study aims to evaluate their health and well-being throughout the transition of the exhibit as well as assess the effectiveness of design features and the associated changes in management practices in achieving program goals.

Data collection began in September 2012, prior to the start of construction, and continued during construction and through the end of 2016, one year after the opening of Elephant Lands. Data collection included regular behavior assessments, adrenal and reproductive hormone analyses in serum and fecal samples, distance walked, and time spent recumbent. Behavior videos were recorded, and serum and fecal samples were collected every week for all elephants. Anklets with GPS and accelerometer units measuring movement and recumbence were worn by two females and two males for 24-hour periods approximately every two weeks. We are currently processing and analyzing these data and plan to report on the results next year.

We will use the study outcomes to evaluate changes in elephant welfare, to determine how effectively we have met our elephant management goals, to identify any additional areas of improvement, to inform possible "fine-tuning" of our habitat, and to inform future elephant exhibit design elsewhere.

Elephant staff have observed a general increase in activity level. The amount of time individual elephants spend grazing and their options for engaging in activities have increased. An individual that previously may have been displaced from an area or activity can now move somewhere else to engage in their activity of choice without being displaced by a herd mate. We are also seeing more behavioral diversity, especially with regard to locomotion and vocalizations. The elephants manipulate their environment more by pulling, pushing, kneeling, stretching, etc., to obtain food items or to move objects for other reasons (including play behavior). The new pool allows for much more water-based activity, and the elephants clearly enjoy swimming and resting in the pool. Our veterinary and curatorial staff have observed improvements in foot health, with increased tread and creases similar to free-ranging elephants, further validated by professional care staff from the Borneo Wildlife Rescue Unit during a visit in May of 2017. All females continued cycling regularly throughout the transition period and in the new exhibit, which provides one indicator of normal reproductive health for the herd.

CONCLUSION

Our vision for Elephant Lands has been based on incorporating elephants' natural behaviors and herd dynamics in the habitat design and on enabling individuals to exhibit a high level of self-determination and choice throughout their day. A pre-established list of a few nonnegotiable "must-have" items helped us make our vision a reality. As it took shape and became the elephants' new home, our new habitat exceeded our expectations. It will allow our elephant family to grow and thrive for generations to come. The Oregon Zoo has pioneered elephant care for decades, and we are proud of the knowledge we have collected, documented, and shared with colleagues around the world. We are committed to continuing this work to promote the best lives possible for the elephants in our care and to continuously advance elephant care on a global level. We are also committed to long-term scientific monitoring of our herd, which will allow us to adjust elements of the habitat as needed to support the expression of natural behaviors and maximize elephant welfare.

Ultimately, we aim for each elephant to have the opportunity to exhibit a full range of natural behaviors while living in a socially stable, multigenerational matrilineal herd that is regularly integrated with bull elephants in a manner that meets or exceeds their biological, social, physiological, and psychological needs. We are immensely grateful for the support of our community, which has allowed us to build Elephant Lands and help advance the welfare of the elephants entrusted to our care.

Table 1

Elephants at the Oregon Zoo When Construction Began for Elephant Lands
(Oldest to Youngest)

ELEPHANT	SEX	ORIGIN	DATE OF BIRTH OR TRANSFER TO OREGON ZOO
Packy	Male	Oregon Zoo	April 14, 1962 (birth)
Tusko	Male	India	~ 1971 (birth) ~ unknown (capture, India) June 19, 2005 (transfer to Oregon Zoo)
Sung Surin	Female	Oregon Zoo	December 26, 1982 (birth)
Rama	Male	Oregon Zoo	April 1, 1983 (birth)

ELEPHANT	SEX	ORIGIN	DATE OF BIRTH OR TRANSFER TO OREGON ZOO
Chendra	Female	Borneo, Malaysia	~ 1993 (birth) ~ April 1996 (capture, Sabah Borneo) November 20, 1999 (transfer to Oregon Zoo)
Rose Tu	Female	Oregon Zoo	August 31, 1994 (birth)
Samudra	Male	Oregon Zoo	August 23, 2008 (birth)
Lily	Female	Oregon Zoo	November 30, 2012 (birth)

REFERENCES

Brown, J. L., Paris, S., Prado-Oviedo, N. A., Meehan, C. L., Hogan, J. N., Morfeld, K. A., & Carlstead, K. (2016). Reproductive health assessment of female elephants in North American zoos and association of husbandry practices with reproductive dysfunction in African elephants (*Loxodonta africana*). *PLOS ONE, 11*(7), e0145673.

Christo, C., Wilkinson, M., & Sheldrick, D. D. (2009). *Walking thunder: In the footsteps of the African elephant.* London, United Kingdom: Merrell.

Douglas-Hamilton, I. (1972). *On the ecology and behaviuor of the African elephant: The elephants of Lake Manyara* (Doctoral dissertation), University of Oxford, United Kingdom.

Eisenberg, J. F., McKay, G. M., & Jainudeen, M. (1971). Reproductive behavior of the Asiatic elephant. *Behavior, 38*(13), 193-225.

Evans, K. E., & Harris, S. (2008). Adolescence in male African elephants, *Loxodonta africana*, and the importance of sociality. *Animal Behavior, 76*(3), 779–787.

Greco, B. J., Meehan, C. L., Miller, L. J., Shepherdson, D. J., Morfeld, K. A., Andrews, J., & Mench, J. A. (2016). Elephant management in North American zoos: environmental enrichment, feeding, exercise, and training. *PLOS ONE, 11*(7), e0152490.

Holdgate, M. R., Meehan, C. L., Hogan, J. N., Miller, L. J., Soltis, J., Andrews, J., & Shepherdson, D. J. (2016). Walking behavior of zoo elephants: Associations between GPS-measured daily walking distances and environmental factors, social factors, and welfare indicators. *PLOS ONE, 11*(7), e0150331.

Maslow, A. H. (1943). A theory of human motivation. *Psychological Review, 50*(4), 370.

McKay, G. M. (1973). *Behavior and ecology of the Asiatic elephant in Southeastern Ceylon* (No. 125). Washington, DC: Smithsonian Institution Press.

Meehan, C. L., & Mench, J. A. (2007). The challenge of challenge: Can problem solving opportunities enhance animal welfare? *Applied Animal Behavior Science, 102*(3), 246–261.

Meehan, C. L., Mench, J. A., Carlstead, K., & Hogan, J. N. (2016). Determining connections between the daily lives of zoo elephants and their welfare: An epidemiological approach. *PLOS ONE, 11*(7), e0158124.

Meller, C. L., Croney, C. C., & Shepherdson, D. (2007). Effects of rubberized flooring on Asian elephant behavior in captivity. *Zoo Biology, 26*(1), 51–61.

Mellor, D. J., Hunt, S., & Gusset, M. (2015). *Caring for wildlife: The world zoo and aquarium animal welfare strategy.* Gland, Switzerland: WAZA Executive Office.

Moss, C. J., & Poole, J. H. (1983). Relationships and social structure of African elephants. In R. A. Hinde (Ed.), *Primate social relationships: An integrated approach* (pp. 315–325). Oxford, United Kingdom: Blackwell.

Rasmussen, L. E. L., & Krishnamurthy, V. (2000). How chemical signals integrate Asian elephant society: The known and the unknown. *Zoo Biology, 19*(5), 405–423.

Shepherdson, D. J. (1999). Environmental enrichment for elephants: Current status and future directions. *Journal of Elephant Managers Association, 10,* 69–77.

Shepherdson, D. J., Carlstead, K., Mellen, J. D., & Seidensticker, J. (1993). The influence of food presentation on the behavior of small cats in confined environments. *Zoo Biology, 12*(2), 203–216.

Shepherdson, D. J., Mellen, J. D., & Hutchins, M. (Eds.). (1999). *Second nature: Environmental enrichment for captive animals.* Washington, DC

Sukumar, R. (1992). *The Asian elephant: Ecology and management.* United Kingdom

Chapter 4

A HOLISTIC APPROACH TO ELEPHANT MANAGEMENT

Integrating Health, Behavior, and Research

Erin Ivory, Corinne Kendall, and Guy Lichty
North Carolina Zoo

Improving elephant welfare requires a holistic approach to management that integrates behavior management, healthcare, and research. To maximize elephant well-being, elephant care, veterinary, and research staff need to work closely to ensure that the effects of management changes, such as diet changes, alterations to social structure, and enrichment, are monitored and quantified carefully. Animal needs are met best when the most up-to-date behavioral management and veterinary practices driven by evidence-based decision-making are used.

The North Carolina Zoo initiated a new holistic management strategy to improve elephant welfare in 2014, which we monitored subsequently using the expertise of three different disciplines. By integrating behavior management, veterinary care, and research, we have been able to track the effects of the management changes on behavior, body conditioning, hormones, and nutrition. Here we will discuss current strategies used to implement change, the way in which we have used research to assess the program, and the effects on elephant management overall. We believe there are many benefits in using a holistic approach in collaborative efforts among management, veterinary medicine, and research.

AN APPROACH TO CHANGE

In 2014, the zoo decided to revamp the elephant program to improve its chances for a successful breeding program, increase variability in access to the habitats and total space available to the elephants, create a staff development program, and implement changes to develop a stronger elephant training and behavioral management program.

Before making any management changes, it was important to assess every aspect of the current program and create a strategy for change that would support the vision overall. It was also important to determine any existing challenges or threats to achieving the vision, such as staff concerns or facility modifications required. The initial assessment yielded many key elements that were needed to support the zoo's vision overall. Small social groupings, over-conditioned elephants, underuse of the facility, and the staff's fear of change were a few of the issues identified (Ivory & Lichty, 2016). Further, there were several underlying behavior issues, both human and elephant, that would need to be mitigated to make progress.

BREAKING THE ROUTINE

Routine is common in elephant management. However, a behavior-driven program that uses trust-based relationships, animal choice and control, and environmental analysis to manage behavior can help minimize routine and improve welfare outcomes (Lacinak, Turner, & Kuczaj, 1995). In this system of behavior management, all aspects of the elephants' daily lives were assessed to address management challenges. Each component of the elephants' experience—habitat, enrichment, training, daily operation, and social management—were evaluated in conjunction with one another because each of the factors is related and none occurs in a vacuum. By looking at the elephants' environment holistically, any physical, mental, behavioral, or social challenge can be mitigated using a combination of management factors to change their environment to improve or modify the behavior. In a behavior-based program, variability, rather than routine, is incredibly important to maintain behaviors desired and extinguish any that are inappropriate or undesired (Lacinak et al., 1995).

SOCIAL STRUCTURE AND BREEDING

Recent research has highlighted the importance of herd integration and diverse social groupings to improve elephant welfare (Greco et al., 2016). Prior to 2014, our herd had not been integrated completely. While all of the elephants had been housed with others at one time or another, in the current management system, the four cows were managed in two social groups, while the two bulls were kept alone for various social and health-related reasons.

There were also staff concerns about herd integration, as perceived personalities attributed to the elephants had heightened the staff's fears of the potential risks (Ivory & Lichty, 2016). Regardless of the management system, the sheer size and strength of elephants can be daunting when contemplating an introduction. Injury or even death is a possibility and, should the introduction begin to deteriorate and aggression ensue, because of the size of the elephants, the keepers cannot intercede physically. However, objective behavior observations, a strong foundation in behavior management principals, and a detailed plan for various potential scenarios can minimize these risks (Turner & Tompkins, 1989).

Elephant hierarchies can be challenging, as each elephant has a different personality, and some tend to be more dominant than others (Laule & Desmond, 1991). If two elephants are competing for the top of the hierarchy, then escalating aggression can occur. Further, elephant social behavior is highly complex, in that they create and establish their own relationships and bonds with others. Again, through behavior observations, keepers can learn which elephants tend to spend time together, the way in which they react to a perceived threat or change in their environment, and their general behavior toward one another. Sometimes these observations demonstrate a tight bond, while at other times, they may show that an elephant has little or no interest in another elephant or even dislikes her. In addition, the complex social structure is not necessarily stable over time. Changes in herd composition, death, illness, and birth all have effects on herd structure (Laule & Desmond, 1991). These observations are critical in understanding the way in which to integrate the herd best and, once the elephants have been introduced, maintain its stability.

Staff fears because of aggression during previous introductions had halted further efforts quickly, and they had not resumed. While the initial introduction may result in chasing and perhaps aggression, it is important to allow the elephants to establish their hierarchy fully (Ramirez, 1999). Stopping and starting introductions repeatedly without allowing the hierarchy to become established may result in escalating aggression. Moreover, interceding to prevent or stop aggression may actually reinforce it and increase its intensity and frequency (Turner & Tompkins, 1989). The way in which the trainers interact with the elephants during an introduction also can affect the success of an introduction and behavior of the elephants.

At the North Carolina Zoo, we carried out each introduction differently depending on the individuals involved and their history and observed behavior. We began the introductions in a controlled setting in the barn, where we provided more space and opened various gates to decrease areas in which an elephant could become trapped. In addition, we provided plenty of hay during the introductions. We believed that increased opportunities to feed cooperatively would give the elephants something to focus on while they adjusted to the new social group. Because these were special circumstances, more hay was provided then the elephants typically would need to decrease competition for food. Thus, we used primary reinforcement to reinforce calm, cooperative behavior in the dominant elephant. Depending on the scenario, the subdominant animal also might be reinforced. However, it is vital to ensure that this does not frustrate the dominant animal and cause her to aggress against the submissive elephant. Herd dynamics are complicated, and reinforcing an animal incorrectly may result in frustration, aggression, decreased trust, or a combination of all three (Turner & Tompkins, 1989).

Although we did observe our dominant female chasing others initially, the episodes were very brief, and we reinforced her throughout for calm, relaxed behavior. We have been able to create social groupings that are more complex by introducing bulls to cows and introducing the two separate cow groups to each other. This has been a critical first step in creating a dynamic herd, which is necessary to develop a natural breeding program. Further, it has improved welfare by increasing space use, exercise, and socialization (Greco et al., 2016).

To date, we have not had any breeding that has resulted in pregnancy. While our potential breeding bull has now had access to two cycling females for over two years, he has not demonstrated the behaviors necessary to produce offspring. We are continuing to assess his behavior and the various environmental factors that may be contributing to the issue. This includes finalizing introductions with the remaining two female elephants and looking at the impact his history may be having on him from a social learning perspective.

FACILITY MODIFICATIONS

Often, we must modify our facilities to accommodate changes in management, social structures, and daily operation. To make progress in developing a more behavior-based management system with complex herd structures, several facility modifications were required. It is extremely important for the keepers to be able to access the elephants in the habitat for stationing, recall, and behavior management. Ideally, a keeper should be able to work with the elephants anywhere in the habitat. In addition to stationing, having areas in which to provide husbandry training allows keepers to maximize the time the elephants are on the main habitat and allows staff to increase the elephants' opportunities for learning. We identified areas surrounding the elephant habitat in which we could create behavior management stations. This would allow us to hold elephants for shifting and separations, reinforce good behavior, train new behaviors in the habitat, and provide easier access to establish stronger recall.

Depending on the herd dynamics, hazardous areas in which subordinate animals can be trapped by those that are more dominant may need to be modified to manage the herd safely in the habitat day and night. We have an area leading to and from the habitat that creates a bottleneck. Not only will a proposed modification relieve the bottleneck, but an additional chute system will be created that will provide a shorter, more direct route to the barn, allowing elephants access to the habitat and the barn during thunderstorms or colder weather. This chute will need to be wide enough for multiple elephants to move between the habitat and the barn and have space to pass one another. Other plans involve creating a more dynamic habitat by using vertical space to encourage overhead foraging and play. In addition to adding more trees to increase shade, we are creating several berms for sleeping in one of our habitats, which will allow all of our elephants to have access to either habitat overnight.

We also have made changes in the way in which we use our habitats. Currently, both habitats are used nightly to increase each elephant's time in the habitat per week. Both chute systems are used to increase variability in access to the habitats to shift elephants or clean the habitat. This allows us to increase the time that visitors can view the elephants while we reset the habitats for the nighttime and has increased the efficiency of our daily operation.

TRAINING AND ENRICHMENT

We have developed a behavioral management program at the zoo that uses operant conditioning with positive reinforcement, behavior-based enrichment, and variability to manage the elephants' behavior to increase their physical, mental, and social opportunities. In addition, we use the Least Reinforcing Scenario (LRS) as a response to incorrect or undesired behaviors (Scarpuzzi, Lacinak, Turner, Tompkins, & Force, 1991), as it has been shown that this can reduce aggression during training sessions. The LRS is essentially a two- to three-second pause following an incorrect behavior, where the trainer is as neutral as possible, therefore neither reinforcing nor punishing the incorrect behavior. The animal's response to the LRS is a behavior where the criteria is to return and maintain a calm, attentive demeanor and should be periodically reinforced.

There are several ways to manage a positive reinforcement training program. At North Carolina, we made slight changes to our training program to increase the elephants' learning capabilities. Routine often features prominently in training programs. However, work with marine mammals, which are similar to elephants in their intelligence and complex social structures, has demonstrated that routines create several issues, including opportunities for behavior breakdown, boredom on the part of both the animal and trainer, frustration, and even aggression as the animal recognizes the pattern of the routine (Lacinak et al., 1995). Similarly, there is evidence in other species that routines are linked to behavioral and physiological stressors that decrease welfare (Bassett & Buchanan-Smith, 2007).

The marine mammal profession pioneered the use of behavior-based positive reinforcement training (Kuczaj & Xitco, 2002). Many of the changes made to the North Carolina Zoo program reflect the years of research that marine mammal trainers have conducted with other extremely intelligent social animals. For example, we have adopted a terminal bridge in which the animal is allowed to terminate the behavior following the sound of a whistle. This allows us to pinpoint the exact moment at which the animal is doing what we want him/her to do and has a direct influence on the precision of communication (Ramirez, 1996). In addition, because food delivery follows the bridge, it decreases the focus on the food compared to the behavior and creates more opportunities to pinpoint different aspects of the behavior, which maintains a higher behavioral criterion. For example, training an elephant to accept footwork requires multiple facets of behavior: the speed at which the animal responds to the cue, the placement of the foot, the correct angle and duration, tolerating having the foot care tools touch the foot, and how still the elephant holds the foot in position. It is important to reinforce each aspect of that behavior at different times. Not only does it create more interest on the animal's part by keeping the session variable, but it helps communicate all of the subtle nuances of the behavior to the animal, thereby increasing clarity and decreasing frustration (Ramirez, 1996). The animal also is unable to predict what will happen next, which decreases his/her ability to manipulate the session. In contrast, if one uses the continuous feed method, the animal may become satiated or solicit more food by exhibiting inappropriate behaviors. For example, if the elephant is presenting a foot and you are using continuous feed, the animal may begin to move her trunk toward the keeper working on the foot. If the trainer responds immediately by increasing the rate at which the food is delivered to keep the trunk busy, the animal then learns that when she wants more food, she only has to move her trunk toward the keeper. Elephants are intelligent and can learn human patterns of behavior and use different strategies to manipulate the session, essentially training the trainer.

The way in which the animal is reinforced is extremely important. If the animal receives the same reinforcement for performing the behavior well or poorly, increased effort has no value. In addition, secondary reinforcers, which are stimuli that have acquired reinforcing properties by being paired with those the animal finds inherently rewarding (Hurley, Scaramozzino, & Long, 1999), are important in a training session because they add variability and build a relationship that extends beyond food. This is critical if the animal is sick and uninterested in food. If she has a strong, trusting relationship with the trainer and secondary reinforcers have been used in the program, it increases the likelihood that she still will participate in the session willingly, even if not food-motivated (Ramirez, 1996). Providing variable ratio and reinforcement variety (VRRV) also is an incredible training tool to increase animal motivation (Lacinak et al., 1995). VRRV is based on the principal that high motivation is achieved when the animal cannot predict the outcome that follows a correct behavior. It has been compared to gamblers playing slot machines, in which the occasional reinforcement of dropping coins increases motivation and anticipation: maybe this next pull will pay off.

In this style of management, the elephants are engaged and thinking throughout their entire session. This provides more mental stimulation and encourages the trainer and elephant to communicate and work together to achieve the behavior. In this way, the elephants build very strong relationships with each individual trainer. The single drawback to this system is that it does require a team of behaviorists that understands how to read the animal's subtle cues and the nuances of behavior and, in turn, understands how to communicate to the animal better what they are seeking (Ramirez, 1996). In a routine-based system, the trainers can be alternated with very little issue because the animal is performing a set pattern each day. However, it places less emphasis on the individual relationship and skill of the trainer. In addition, the variable training schedule allows us to maximize our training time to address specific issues with each individual elephant.

It is very important to elephants' health and welfare to have an environment that is challenging, both physically and mentally. If used correctly, enrichment is another excellent tool with which to increase variability and target the specific behaviors you wish to increase (Kuczaj et al., 2002). It is critical for elephant care professionals to ensure that enrichment is used in conjunction with operant conditioning and variability (Lacinak et al., 1995). It can be very difficult to move and change elephant enrichment because it usually is heavy and requires large machinery. As a result, it is easy for an elephant enrichment program to become stagnant and routine, thus losing its value over time. It is becoming a common practice to use automatic feeders, and while there definitely are benefits to being able to allocate food on a variable schedule throughout the night, it can reinforce stereotypic behavior or aggression unintentionally (Lacinak et al., 1995). For example, an elephant may be engaged in swaying or chasing another animal when the feeder dispenses food, in which case the elephant will be reinforced for an undesirable behavior, thereby increasing the likelihood that the behavior will occur again. Thus, it is important to analyze all aspects of the care we provide to ensure that we are not creating new issues while trying to solve others.

At North Carolina Zoo, we use a variety of enrichment designed to encourage species-specific behaviors. While we have a more natural habitat where the elephants have access to various mud wallows, pools, browse, scratching trees, dusting piles and grass, we use more specific devices to encourage those natural behaviors in the barn and paddock areas. Since elephants generally demonstrate the most stereotypic behavior in anticipation of routine keeper-driven environmental change or when confined to smaller spaces, we use species-specific environmental enrichment devices during these periods to create more foraging opportunities and more complex foraging experiences while decreasing stereotypy. During the winter months, we also continue to provide opportunities for socialization with other elephants, which can help reduce stereotypic behaviors. The focus of our enrichment program is to create an environment that encourages the behaviors we are looking for (i.e., social interaction, play, foraging, skin care, and physical or mental challenges). We are continuously assessing and are modifying our enrichment approach based on the behavior of the elephants.

Behavior management examines the behavior and the environment in which it occurs, and both are managed with changes made to the environment. Rather than using routines and checklists to ensure care, the team is trained to understand behavior, the way trainer behavior affects elephant behavior, the benefits of variability, and ways to create a dynamic habitat to increase species-specific behaviors and decrease inappropriate and stereotypic behaviors.

DEVELOPING A BEHAVIOR-BASED STAFF DEVELOPMENT PROGRAM

To change the elephant management program, the entire staff had to be trained in behavior management. This is an ongoing process, and it will require several years before all of the keepers are fully proficient. We created a development program based on a variety of job-related proficiencies to allow each keeper, regardless of title or tenure, to move through the various levels based on his/her individual skills and ability to grasp each proficiency. This program is not based on how long a keeper has been in the program or at the zoo, but on the skills demonstrated. Each level requires supplemental training courses in operant conditioning, management, health care, and reproduction to increase each staff member's knowledge. The development program consists of five levels that all must achieve, regardless of prior experience, ranging from day one in the elephant barn through senior elephant care specialist. This elephant management system requires a different set of skills with which certain keepers may struggle. On the other hand, staff members who demonstrate high behavioral proficiency and strong leadership qualities are eligible to participate in our international projects that focus on elephant welfare and conservation.

STRATEGY FOR CHANGE

The concept of successive approximations is one of the cornerstones of learning for all animals, including humans. Successive approximation entails breaking a behavior down into tiny steps that are trained incrementally until the final behavior is mastered (Hurley et al., 1999). Managing change through successive approximations involves looking at the big picture of all the elements needed for a strong elephant management program and breaking down each component into small, digestible pieces for the staff. As the staff and the elephants become adjusted increasingly to change and develop trust, these approximations can occur in closer succession, and more challenging changes can be attempted.

Every change made in the barn, whether it was operational (how the barn is cleaned), management (variability of elephant social structures), or behavioral (using an LRS as a response for incorrect behavior), was implemented incrementally to gain the team's acceptance and allow them the opportunity to understand the process of change. This was a key component in the staff's development. Fear can be a good deterrent to impulsive action; however, addressing those fears and mitigating them is an extremely valuable set of skills that will allow the program and the team to conquer those fears and ultimately allow the elephants to lead more complex and enriching lives. At North Carolina Zoo, we needed to challenge the elephants incrementally to discover what limitations each elephant actually possesses compared to what limitations the staff perceived. We accomplished this through small changes in their daily care and creating scenarios that allow the animals to choose whether they would like to proceed. We use positive reinforcement to encourage the behaviors we want to see, such as shifting to a new area or being in closer proximity to another elephant. Throughout the entire process, we discussed the elephants' behavior and worked as a team to set the next goal or priority.

INITIAL ANECDOTAL RESULTS

Over the past two years, the staff has made anecdotal observations of changes in the elephants' behavior and demeanor. In general, we have seen a decrease in aggression, both that directed at humans and one another. The elephants seem more relaxed and calmer and have been far more engaged in the training sessions, demonstrated by their willingness to participate, their faster responses, and general demeanor during the sessions. The elephants also have been observed interacting and displaying more social behavior than before. In particular, one of our males and females play every day around lunchtime for an extended period. The introductions yielded surprising results, as one of our less socialized elephants, whose potential reaction to the introduction created the most concern among staff, not only handled the introductions very well, but also acted as a mediator during conflict by placing herself between two elephants and defusing the situation.

All of these behavioral changes in the elephants have encouraged staff to continue modifying our behavioral management program. The team members have become more comfortable and willing to work through complex behavior challenges and are active participants in continuing to develop the program. Further, there has been a marked shift in the way they perceive the elephants, and they work to ensure that they do not label them. Productive discussions now occur frequently to analyze the behavior within the context of the environment.

PROGRAM ASSESSMENT: AN ADAPTIVE MANAGEMENT APPROACH

We need comprehensive, hypothesis-driven management manipulations to improve knowledge about elephant welfare (Barber, 2009). Recent multi-institutional studies have provided valuable information about the correlations between various physical exhibit factors and elephant welfare outcomes (Meehan, Mench, Carlstead, & Hogan, 2016). However, these studies have lacked the experimental approach necessary to understand the way management changes can alter elephant well-being and must operate within the confines of existing management approaches, which may limit their use in truly understanding elephant welfare (Barber, 2009). Ongoing improvements in management offer a unique opportunity to assess the way changes in husbandry affect elephant well-being.

Using an adaptive management approach, management changes can be treated as experiments. By gathering data before and after different changes, the effects of these alterations can be measured, and we can verify whether they have had the desired outcome. Once analyzed, this research can then be used to inform whether or not the changes should be maintained and if they have met the goals necessary for different measures of well-being (e.g., a body condition score of 3, at least five hours spent walking daily, or performance of repetitive behaviors less than 5% of the time). Even if management changes do lead to positive outcomes, it may be necessary to make additional modifications to reach an outcome desired. By integrating management with research, it becomes possible to go beyond rudimentary assessments of what we provide to assess elephant well-being outcomes themselves. An adaptive management approach ensures that institutions are assessing and improving their programs constantly using an evidence-based approach.

A holistic approach to elephant management that integrates research and care has several other advantages. First, an adaptive management approach in which research is incorporated in management practices ensures that maximal information is captured as changes are made. While it is difficult to conduct controlled experiments in zoo settings, this approach can help provide valuable information about the way husbandry changes alter behavior, reproductive outcomes, or health, which is needed as an alternative to "fishing expedition" surveys, which provide only correlational data (Barber, 2009). In addition, such an approach provides managers with data to support husbandry changes that may be staff intensive or contentious. Through ongoing communication between keepers and researchers, staff obtain valuable insights about the outcomes of these changes, and managers can thus garner greater staff support for ongoing efforts.

APPLYING RESEARCH TO ELEPHANT MANAGEMENT: EXAMPLES FROM NORTH CAROLINA ZOO'S OBJECTIVES

In a holistic management program, the primary research objective is to assess the effects of management modifications. By gathering data before and after major changes in social structure, diet, facility modifications, or enrichment, it is possible to begin to determine the way management affects behavior, nutrition, reproduction, and health. Although some changes may require a long while to have an effect, measuring the effects of change is an important first step in ensuring that management decisions have the effect desired. Establishing simple, ongoing data collection can help ensure that the effects of change are quantifiable, even if only in the long term.

Many research methods already have been developed to assess physical and mental health in elephants. By using these existing methods and continuing to build on them, it is possible to measure key physical and mental health outcomes. In this section, we provide examples of simple research methods that we have used in combination with management changes to monitor the influence of our husbandry decisions.

DIET AND BODY CONDITION

Zoos increasingly are moving to reduced calorie, grain-free diets that approximate natural diets more closely (Less, Lukas, et al., 2014). In addition, by shifting from small quantities of high-calorie foods to large quantities of low-calorie foods, zoos can give animals more opportunities to engage in natural foraging behaviors (Bergl et al., 2014). Recent studies have demonstrated that over-conditioning is a problem in captive elephants (Morfeld, Meehan, Hogan, & Brown, 2016). Elephants often display stereotypic and anticipatory behaviors, and studies suggest that unpredictability in diet provision, more food provisioning times, and use of low-starch diets may help reduce these undesirable behaviors (Morfeld et al., 2016; Less et al., 2014).

North Carolina Zoo made changes to the elephant diet in 2016 to improve welfare. These shifts have involved use of a new hay balancer pellet that is nutritionally complete but has fewer calories and is produced by Mazuri, an exotic animal food company. The elephant diet now consists primarily of hay, and ongoing efforts are being made to increase the amount of browse provided as well. In addition, we have made efforts to reduce the use of sugary rewards, such as sweet feed or molasses, and shift to lower-calorie produce such as carrots. The zoo is assessing body condition to monitor the effects of this diet change. Building on previous work that relied on body condition scores alone, which can be subjective, deuterium, a stable isotope of water, is used to estimate the elephants' percent body fat. Deuterium is administered to the elephant in a known quantity, and then the ratio of deuterium to water is measured.

Similar techniques have been used successfully in several other species (Dugdale, Curtis, Cripps, Harris, & Argo, 2010; Garcia, Rosetta, Ancel, Lee, & Caloin, 2004; Mawby et al., 2004). In this manner, the animal's water volume can be calculated, and from this, percent body fat is estimated. We used this technique to assess each elephant's body condition before and after the diet change. This study is ongoing, and body condition assessment is conducted every six months. This study will allow the zoo to determine whether the desired outcome of reduced body fat is being achieved, rather than just reduced weight or body condition scores.

This technique has been particularly valuable for older animals that are part of the study. One of the bull elephants is over 40 years of age and suffers from arthritis and cataracts, which limit his mobility. This bull's diet was increased after a prolonged illness and led to considerable weight gain. Initial weight gain was desirable, but now there are concerns that the animal may be over-conditioned. However, because of his condition and age, it is important to ensure that weight loss that might occur with diet changes is attributable to loss of fat, not muscle. Use of deuterium to assess changes in fat composition allows the zoo to monitor changes and is more specific than weight monitoring alone, which is critical to the health outcomes desired for this individual.

In addition to changes in body condition, the zoo also is monitoring the nutritional outcomes of diet changes in collaboration with a master's student from North Carolina State University. These have involved micronutrient analysis of elephant blood samples before and after diet changes. A long-term goal is to shift the elephants entirely to a browse and hay diet, with limited produce and vitamin supplementation. To accomplish this, it is important to understand the micronutrients that different browse species contain. We conduct careful micronutrient analyses of various tree species and parts of the plant (leaves, branches, etc.) and hay every six weeks to establish whether browse can be used in future to replace the hay balancer, which serves currently primarily as a vitamin supplement. By assessing changes in micronutrients of the browse and availability throughout the year, we also are able to assess whether browse could be used as a supplement year-round or only during certain seasons.

CYCLING

Acyclicity is known to be a major issue in captive elephants (Brown et al., 2016). Therefore, monitoring female elephants' hormone profiles is an important part of their regular healthcare. Several factors, including stress, body condition, and social structure, can affect acyclicity, but the effects of these are not yet understood well (Brown et al., 2016). Recent hormone analysis conducted on our elephants suggest that the female dominance hierarchy also may play an important role and suppress regular cycling in some individuals (this study is discussed in more detail in chapter 10). As part of our breeding program, we have implemented introductions between male and female elephants, as well as increased diversity in social groupings. Monitoring hormone profiles regularly throughout this process is critical in understanding the effect that such social change has on elephant reproductive health.

BEHAVIOR

As intelligent animals who naturally live in complex social groups, zoo elephants are prone to repetitive behaviors, such as rocking and pacing (Greco et al., 2016). Baseline data on elephant activity budgets, how time of day influences them, and individual differences in behavior are critical to assess management changes designed to reduce undesirable behaviors. The zoo has conducted behavioral research on its elephants periodically for several years. Hasenjager and Bergl (2015) provided important baseline data on elephant behavior at the zoo, and their ethogram has been adopted for further use (see Table 1). With any significant management changes, it is key to monitor behavior to assess whether or not desired outcomes are being achieved. In recent years, North Carolina has made changes in herd structure, as well as the use of browse and enrichment, in efforts to improve elephant well-being. To assess these changes, the zoo implemented a behavioral monitoring program beginning in 2015. This program uses a simple ethogram and scan sampling to determine changes in each elephant's activity budget over time. Because zoo staff resources for such data collection are limited, the zoo created an internship program in collaboration with North Carolina State University. This program uses undergraduate students who are able to collect data while the elephants are on exhibit during the day, as well as record behaviors during the winter and overnight from video monitoring in the barn. This research provides vital information on the way various management changes affect elephant behavior.

Similarly, the zoo has used behavioral research to address concerns about exhibit design and shade availability. We have been working to assess whether we need more shade structures to ensure elephants have adequate protection from the elements. To address this concern, the behavior monitoring program now includes information on elephant shade use throughout the day and over the course of the year. Using mapping models, it also has been possible to calculate how shade availability changes with time of day and season. By combining information on shade availability in the exhibit, as well as the animals' actual shade use, it is possible to assess whether shade availability is an issue. Similarly, for institutions concerned about space use or elephant movement, behavioral monitoring in combination with GPS or accelerometer use can provide valuable insights about what facility modifications might be needed (Rothwell, Bercovitch, Andrews, & Anderson, 2011).

CONCLUSION

All zoos should make efforts to use a management style that increases animals' opportunity for choice and uses trust-based relationships and positive reinforcement training to improve elephant well-being. By using a behavior management approach, one can assess the environment in conjunction with the behavior of the individual and group and, thus, use variability in the environment to achieve the changes in behavior desired. A holistic approach to management ensures that all aspects of elephant welfare are considered when creating management goals. By integrating resources from the veterinary, research, and husbandry departments, it is possible to use a holistic management approach that maximizes animal welfare. Management efforts need to be proactive rather than reactive, but by assessing the outcome of management decisions, it is possible to monitor well-being outcomes, improve staff engagement, and increase scientific knowledge about elephant welfare.

Table 1

Ethogram of African Elephant Behavior (Adapted from Hasenjager & Bergl, 2015)

BEHAVIOR	DESCRIPTION
Self-maintenance	Rubbing, scratching, throwing sand/dirt/hay/water/fecal material onto body, mudding, dusting, or digging
Resting Standing	No interactions with conspecifics, environment, or keepers; no bodily movement. Can be asleep or awake while standing
Resting Lying down	No interactions with conspecifics, environment, or keepers; no bodily movement. Can be asleep or awake and is lying down
Resting Partial	No interactions with conspecifics, environment, or keepers; no bodily movement. Can be asleep or awake and is partially lying down. Head may still be upright, but body is touching ground
Resting Stretch	No interactions with conspecifics, environment, or keepers; no bodily movement. Elephant supports itself on its knees and elbows with rear legs extended back and front legs extended forward

Pacing	Pacing back and forth between point A and B or in a circle. May include eating
BEHAVIOR	**DESCRIPTION**
Rocking	Swaying back and forth with no locomotion. May include eating
Other repetitive behavior	Other repetitive behavior whose cause and function are unknown. May include eating
Locomotion	Walking or running nonrepetitively
Spinning	Elephant turns in circle repetitively
Vocalization	Elephant makes an audible sound
Forage Exhibit vegetation	Browsing on trees/shrubs/grass
Forage Prepared diet	Eating prepared diet (hay/grain/vegetables)
Forage Browse	Eating cut browse
Forage Sand	Eating sand.
Drinking	Drinking from pool or drinker
Interacting Exhibit structure	Interacting with permanent exhibit structures. Does not include eating
Interacting Caging	Interacting with cage bars/fencing, both on- and off-exhibit

Interacting Temporary Enrichment	Manipulating any temporary nonfood enrichment item, such as tires or barrels
BEHAVIOR	**DESCRIPTION**
Bathe	Bathing in exhibit pond. Does not include splashing water while standing at the pool edge
Other Solitary Behavior	Other solitary behaviors (e.g., moving trunk, defecating, etc.)
Agonism Non-contact	Threats (ears extended, charging, head shake, pursuit). No physical contact between individuals
Agonism Contact	Aggressive behavior involving physical contact between individuals; can include biting, head-butting, poking, striking with the trunk, or pushes
Affiliative Contact	Behaviors that involve any nonaggressive physical contact; includes trunk intertwining, trunk placed within another elephant's mouth, contact with another elephant without obvious use of force
Displace	Focal elephant approaches another elephant which then leaves its position; the focal elephant takes up the vacant position
Displaced	Focal elephant moves from its position as another elephant approaches it; the other elephant then takes up the vacant position
Keeper interaction	Interacting with a keeper
Not visible	Elephant or its behavior is not visible to the observer
Other	Elephant engages in any behavior that does not meet the above behaviors

REFERENCES

Atkinson, S. N., Nelson, R. A., & Ramsay, M. A. (1996). Changes in the body composition of fasting polar bears (*Ursus maritimus*): The effect of relative fatness on protein conservation. *Physiological Zoology, 69*(2), 304–316.

Barber, J. C. (2009). Unpacking the trunk: Using basic research approaches to identify and address captive elephant welfare concerns. In D. L. Forthman, L. F. Kane, D. Hancocks, & P. F. Waldau (Eds.), *An elephant in the room: The science and well-being of elephants in captivity* (pp. pp. 111–128). North Grafton, MA: Tufts University Press.

Bassett, L., & Buchanan-Smith, H. M. (2007). Effects of predictability on the welfare of captive animals. *Applied Animal Behaviour Science, 102*(3), 223–245. Retrieved from http://dx.doi.org/10.1016/j.applanim.2006.05.029.

Bergl, R., Ball, R., Dennis, P. M., Kuhar, C. W., Lavin, S. R., Raghanti, M. A., & Lukas, K. E. (2014). Implementing a low-starch biscuit-free diet in zoo gorillas: The impact on behavior. *Zoo Biology, 33*(1), 63–73.

Brown, J. L., Paris, S., Prado-Oviedo, N. A., Meehan, C. L., Hogan, J. N., Morfeld, K. A., & Carlstead, K. (2016). Reproductive health assessment of female elephants in North American zoos and association of husbandry practices with reproductive dysfunction in African elephants (*Loxodonta africana*). *PLOS ONE, 11*(7), e0145673. Retrieved from http://dx.doi.org/10.1371/journal.pone.0145673.

Dugdale, A. H. A., Curtis, G. C., Cripps, P., Harris, P. A., & Argo, C. M. (2010). Effect of dietary restriction on body condition, composition, and welfare of overweight and obese pony mares. *Equine Veterinary Journal, 4*(7), 600–610. Retrieved from http://dx.doi.org/10.1111/j.2042-3306.2010.00110.x.

Garcia, C., Rosetta, L., Ancel, A., Lee, P. C., & Caloin, M. (2004). Kinetics of stable isotope and body composition in olive baboons (*Papio anubis*) estimated by deuterium dilution space: A pilot study. *Journal of Medical Primatology, 33*(3), 146–151. Retrieved from http://dx.doi.org/10.1111/j.1600-0684.2004.00064.x.

Greco, B. J., Meehan, C. L., Hogan, J. N., Leighty, K. A., Mellen, J., Mason, G. J., & Mench, J. A. (2016). The days and nights of zoo elephants: Using epidemiology to better understand stereotypic behavior of African elephants (*Loxodonta africana*) and Asian elephants (*Elephas maximus*) in North American zoos. *PLOS ONE, 11*(7), e0144276. Retrieved from http://dx.doi.org/10.1371/journal.pone.0144276.

Hasenjager, M. J., & Bergl, R. A. (2015). Environmental conditions associated with repetitive behavior in a group of African elephants. *Zoo Biology, 34*(3), 201–210. Retrieved from http://dx.doi.org/10.1002/zoo.21211.

Hurley, J. A., Scaramozzino, J. M., & Long, J. M. (1999). Training and behavior terms glossary for the International Marine Animal Trainers Association (IMATA). In K. Ramirez (Ed.), *Animal training: Successful animal management through positive reinforcement* (pp. 164). Chicago, IL: Shedd Aquarium Press.

Ivory, E., & Lichty, G. (2016). The potential pitfalls of risk avoidance in elephant management. *Proceedings of the Association of Zoos and Aquariums Conference,* San Diego.

Kuczaj, S., Lacinak, T., Fad, O., Trone, M., Solangi, M., & Ramos, J. (2002). Keeping environmental enrichment enriching. *International Journal of Comparative Psychology,* 15(2), 127-137.

Kuczaj, S. A., & Xitco, M. J., Jr. (2002). It takes more than fish: The psychology of marine mammal training. *International Journal of Comparative Psychology,* 15(2), 186-200.

Lacinak, C. T., Turner, T. N., & Kuczaj, S. A. (1995). Training, enrichment, and behavior. *Proceedings of the Association of Zoos and Aquariums Conference, Seattle.*

Laule, G., & Desmond, T. (1991). Meeting behavioral objectives while maintaining healthy social behavior and dominance: A delicate balance. *Proceedings of the International Marine Mammal Trainers Association Conference, Vallejo, CA.*

Mawby, D. I., Bartges, J. W., d'Avignon, A., Laflamme, D. P., Moyers, T. D., & Cottrell, T. (2004). Comparison of various methods for estimating body fat in dogs. *Journal of the American Animal Hospital Association,* 40(2), 109–114.

Meehan, C. L., Mench, J. A., Carlstead, K., & Hogan, J. N. (2016). Determining connections between the daily lives of zoo elephants and their welfare: An epidemiological approach. *PLOS ONE, 11*(7), e0158124. Retrieved from http://dx.doi.org/10.1371/journal.pone.0158124.

Morfeld, K. A., Meehan, C. L., Hogan, J. N., & Brown, J. L. (2016). Assessment of body condition in African (*Loxodonta africana*) and Asian (*Elephas maximus*) elephants in North American zoos and management practices associated with high body condition scores. *PLOS ONE, 11*(7), e0155146. Retrieved from http://dx.doi.org/10.1371/journal.pone.0155146.

Ramirez, K. (1996). Secondary reinforcers as an indispensable tool: The effectiveness of non-food reinforcers. *Marine Mammals: Public Display and Research, 2*(1).

Ramirez, K. (1999). *Animal training: Successful animal management through positive reinforcement.* Chicago, IL: Shedd Aquarium Press.

Rothwell, E. S., Bercovitch, F. B., Andrews, J. R., & Anderson, M. J. (2011). Estimating daily walking distance of captive African elephants using an accelerometer. *Zoo Biology, 30*(5), 579–591.

Scarpuzzi, M. R., Lacinak, C. T., Turner, T. N., Tompkins, C. D., & Force, D. L. (1991). Decreasing the frequency of behavior through extinction: An application for the training of marine mammals. *Proceedings of the International Marine Animal Trainers Association, Amsterdam.*

Turner, T. N., & Tompkins, C. D. (1989). Aggression: Exploring the causes and possible reduction techniques. *Proceedings of the International Marine Animal Trainer Association, Amsterdam.*

Chapter 5

USING BEHAVIOR AND HEALTH TO MONITOR WELLNESS AND EVALUATE THE INFLUENCE OF EXHIBIT DESIGN ON AFRICAN ELEPHANTS AT CLEVELAND METROPARKS ZOO

Chris D. Peterson
Associate Curator of Animals/Elephant Manager

Bonnie A. Baird, PhD[1]
Graduate Research Associate

Cleveland Metroparks Zoo, Cleveland, Ohio, USA

INTRODUCTION

Cleveland Metroparks Zoo's (CMZ) excellence in elephant care is characterized by a program that focuses on three key aspects of elephant management: creating and maintaining a cohesive team, managing to maximize elephant wellness, and using science to evaluate and inform practices continually. The mission of CMZ's Elephant Care Team is to increase the knowledge and understanding of elephants and to convey this knowledge to our visitors, staff, volunteers, and peers. We aspire to have a unified team that supports elephant conservation and research,

---

1 Current Affiliation: Animal Welfare Scientist, Woodland Park Zoo, Seattle, WA, USA

provides a physically and socially fulfilling environment for the elephants, and ensures staff and guest safety. Our elephant management philosophy focuses on achieving the highest possible level of welfare for the elephants in our care by providing opportunities to display natural behaviors, continually enhancing habitats, managing to respect natural herd dynamics, and using innovative husbandry and training techniques that prioritize elephant health and wellness. By using team and science based management, we not only meet the expectations listed above, but our program emphasizes continual empirical evaluation and improvement of our practices.

THE ELEPHANT CARE TEAM

CMZ's Zoological Programs Division encompasses our Animal Programs, Veterinary Programs, and Conservation and Science teams. Our success is attributable to a multidisciplinary approach that values each team's skills equally. The Animal Programs team, including curators and animal keepers, is primarily responsible for direct care, knowledge of the individual animals, observations, and enacting program goals. The Veterinary Programs team includes the veterinarians and technicians who provide medical care and also are essential participants in the training processes that help ensure compliance in future treatments. Veterinary staff dedicate time each week to perform rounds at the elephant facility to maintain engagement with staff and elephants. Through applied research in endocrinology, epidemiology, and behavior, the Conservation and Science team are invaluable in documenting and validating keeper observations, challenges, and program adjustments. The Elephant Care Team members help one another achieve the elephant management program's mission through teamwork, respect, trust, communication, problem solving, and accountability.

ANIMAL PROGRAMS

A clear elephant management philosophy, strong leadership, and a team based focus help ensure successful implementation of CMZ's elephant management program. Although the Associate Curator of Animals/ Elephant Manager is responsible for daily oversight of the program, the program's success is largely keeper driven. It is critical to have experienced keepers who believe in the program and place the animals' best interests foremost. Every action an animal keeper performs is for the elephants' benefit, rather than convenience. Despite the fact that this inevitably adds effort and labor costs, the elephant husbandry staff has made a commitment to perform those duties in a manner that benefits elephant care. A tangible example is setting up multiple feeding stations, rather than having the convenience of one, to promote exercise and enhance herd dynamics.

Of equal importance is the tiered staffing, which emphasizes development and succession. The objective is to assemble a staff that supports a perpetually evolving program. Animal keepers are assessed and their proficiencies ranked bi-annually. A scale of 1 through 5 was developed to describe each keeper's experience level. Level One keepers know the CMZ Elephant Management Guidelines and have basic elephant knowledge. Level Two keepers have a basic understanding of training techniques and can perform routine husbandry tasks (e.g., feeding, brushing feet). To attain Level Three, a keeper must be familiar with the daily operation of the program and be able to act as the leader of the day. A benchmark of a Level Three keeper is the ability to work an elephant through a bath routine. Level Four keepers demonstrate increased proficiency with respect to elephant training and knowledge. In addition, they have a working understanding of all husbandry techniques (e.g., foot trimming, trunk washes, and restraint). Achieving Level Five requires a keeper to be proficient in all areas of elephant care and husbandry. She/ he also must demonstrate good leadership, serve as a mentor, and be an advocate for CMZ's elephant program.

Day-to-day, the keepers function as one unit, rather than as individuals. There are team meetings each morning to review the previous day's occurrences and set the current day's goals for the elephants and their habitats. The keeper responsible for the day's documentation takes minutes following a standardized template, which are later reviewed by keepers working on subsequent days. The types of activities documented include notable events, elephant husbandry (e.g., foot care, baths, and veterinary examinations) and training goals, and enrichment strategies. If advanced training sessions (e.g., restraint training) are planned, keepers review notes from the previous training session. They also discuss the day's exhibitry plans and coordinate guest experience opportunities in conjunction with Conservation Education staff. Each team member's roles and responsibilities are defined in all of these goals, with the understanding that roles may flex to meet the animals ever changing needs. The core of our team framework is not only to coordinate individual responsibilities, but also to take advantage of opportunities to work as a group.

When the team performs husbandry interventions, members assemble before a session begins and again after it concludes. The pre-meeting clearly defines the session's purpose and responsibilities of each team member. The post-meeting provides real-time feedback to each team member about observations of both the primary keepers and the elephant. The pre- and post-session assemblies support an environment in which team members engage in open and constructive guidance. Sessions also are video recorded for review. Recorded sessions are useful for both instruction and to evaluate occurrences of abnormal elephant behavior.

Further, our team includes the subjective input of an elephant care consultant to supplement the experience of CMZ's team. Recognizing that staff must focus on the events that take place at the zoo in which they work, a consultant often can provide guidance based on a wider range of experiences. The outside perspective is less biased in evaluating the zoo's efforts and can communicate industry wide challenges or standards that translate into goals for improvement.

In addition to keeper proficiency and a unified philosophy, trust is a major component of success. Curators trust the keepers to care for the elephants efficiently and effectively, while keepers trust that curators have the best interests of the elephants, staff, and Zoo in mind when making decisions.

VETERINARY PROGRAMS

The Veterinary staff's frequent presence at the elephant facility is an invaluable component of our program. Under protected contact, a brief, weekly physical examination of each animal is performed that includes a hands-on inspection of the elephant's feet, eyes, oral cavity and teeth, body condition, and gait. A voluntary blood draw from the ear vein usually concludes the exam. Both the clinicians and Veterinary Technicians are involved in the rotation so that the entire Veterinary staff is able to perform any procedure on any elephant worked by any keeper/trainer. The close working relationships that develop between the elephant keepers/trainers, Veterinary staff, and the elephants themselves when performing these tasks on a regular basis promote dialogue and trust that has paid dividends when health challenges arise. The commitment of personnel, time, resources, and effort was a conscious decision by Animal Programs and the Veterinary staff as a testament to the importance of our Elephant Management Program.

The Veterinary staff ensures that blood samples from each elephant are analyzed quarterly for routine bloodwork (CBC/differential and chemistry profiles) to monitor their health overall. The Veterinary Technicians bank the serum from the weekly phlebotomies in an ultra-low temperature freezer as a research tool and for retrospective diagnostics as necessary. In addition, having a full endocrinology lab in the hospital allows us to measure physiological parameters over time to quantify and evaluate how management changes affect the elephants. Ready access to assess long-term trends allows us to develop a more complete picture of the health and welfare of these animals.

CONSERVATION & SCIENCE

Using science to inform husbandry decisions is an increasingly essential component of animal management in zoos. Although exotic animals have been housed in captivity for centuries, relatively few investigations have empirically evaluated welfare in these species, and the studies that have been carried out in accredited zoological institutions are biased heavily toward mammals, particularly primates (Stoinski, Lukas, & Maple, 1998; ; Melfi, 2009). This lack of empirical information leads to a general, "one size fits all" approach to housing and husbandry guidelines in many species that often is based on historical traditions (Melfi, 2009). However, without empirical evaluations of these traditions, we have no way of knowing their true influence on animal well-being. Many husbandry practices are used regularly with a variety of species and are thought to have a neutral effect on animal welfare, but critical evaluation may reveal that a small change in practice would result in vast improvements in animal wellness overall (e.g., providing hiding spaces for captive small felids (Shepardson. Carlstead, & Wielebnowski 2004; Carlstead Brown, & Seidensticker, 1993). The idea that housing and husbandry decisions should be based on evidence rather than tradition whenever possible is the basic principle underlying science-based zoo management (Melfi, 2009).

In pursuit of that goal, CMZ is committed to integrating our science and animal care teams seamlessly (Kuhar, Kornak, & Lukas , 2015). The Zoo has a world-class research program on site, with five PhDs on staff, a full endocrinology lab, and an extensive graduate program. Our partnership with Case Western Reserve University allows up to four PhD candidates at a time to be trained at the Zoo in a unique program that integrates research, conservation, and zoo leadership. Partnerships with other local universities allow Master's level students to conduct their thesis research under the guidance of zoo professionals as well. Students at every level are integrated into daily operations and work closely with animal care teams and zoo leadership. This collaboration allows for open discussions and sharing of ideas among employees in different roles that benefit the animals in our care. Constant communication is also crucial for success in every phase of the research process, from study design to integrating results into daily husbandry practices.

The Zoo's Endocrinology Lab also helps inform husbandry and management decisions by monitoring reproductive hormones in all five elephants regularly. The Endocrine Lab Manager analyzes the serum samples collected weekly by the Veterinary Team, and circulates monthly reports to the rest of the Elephant Care Team so changes can be monitored and correlated with the behavioral observations the latter make. Progesterone levels indicate ovarian cyclicity, which may be related to health and welfare in elephants (see Brown et al., 2016). In addition, the male's testosterone levels are measured, which allows physiological confirmation of any behavioral signs of musth observed by the Animal Care team.

MANAGING FOR HERD DYNAMICS

Elephants are incredibly social creatures with complex relationships that play a large part in their daily lives. Consequently, herd dynamics are at the heart of all elephant husbandry and management decisions. Ideally, all such actions should strive to strengthen herd dynamics and avoid degrading herd structure. Habitat management, elephant shifting, training, and evaluation are all crucial components of managing herd dynamics. In 2011, CMZ opened African Elephant Crossing (AEC), a dynamic new exhibit that consists of multiple elephant spaces that set the stage for a more committed approach to managing herd dynamics.

HABITAT MANAGEMENT

Four diverse elephant habitats include a sheltered interior space (Fig 1), a flat, open Savanna exhibit (Fig 2), a tree-filled "Mopani yard" exhibit (Fig 3), and an off-exhibit exterior "Night" yard (Fig 4). These spaces provide varied opportunities for shelter, shade, supplemental heat, substrates, water features, and terrain. The interior space is the primary source of shelter; however, there also are mature trees and covered areas with overhead heat in the outdoor areas. There are heated floors throughout the inside, as well as some areas outside the facility. The flooring/substrate options range from the traditional smooth finished floors to rubberized coated floors, sand stalls, and natural substrate over variable terrain. The water feature options range from zero entry to full submersion. Habitats at CMZ are designed to afford the elephants opportunities to socialize, roam, forage, wallow, dust, and bathe, and promoting these natural behaviors is the basis for our elephant exercise and enrichment program. The management philosophy is to rotate the elephants among these spaces throughout the day to expose each to as much variability as possible. Each shift into a new habitat requires the yard to be prepared with new feeding choices, enrichment activities, and social stimulation. Further, an "elephant crossing" guest experience is incorporated in these transfers. Twice a day, this "elephant crossing" offers us the opportunity to explain our purpose and philosophies of elephant management and conservation to our guests, in addition to moving the elephants into a different habitat.

Fig. 1: The interior space features heated floors, elevated feeders, and a mix of smooth and sand floors.

FIG. 2: THE SAVANNA EXHIBIT FEATURES TWO POOLS AND IS FLAT AND OPEN TO RESEMBLE THE AFRICAN SAVANNA (**A**). ELEVATED FEEDERS AND A SHADE STRUCTURE WERE ADDED IN 2015 (**B**).

FIG. 3: THE MOPANI YARD FEATURES AN ELEVATION CHANGE, TREES
FOR SHADE (**A**), A LARGE MUD WALLOW, AND A PARTIALLY BURIED
CONCRETE CULVERT WITH HOLES FOR HIDING FOOD ITEMS (**B**).

FIG. 4: THE OFF-EXHIBIT NIGHT YARD FEATURES A HEATED CONCRETE PAD TO ALLOW FOR ACCESS DURING COLDER MONTHS, ELEVATED FEEDERS, AND SAND SUBSTRATE.

Since AEC opened in 2011, science and acute observations have enabled continual exhibit enhancements. The most noteworthy is the addition of seventeen elevated feeders that were placed strategically throughout exhibit spaces, both indoors and outdoors. Locations were chosen to promote movement throughout the exhibit, as well as give the elephants a choice of feeding locations. Electric winches equipped with hooks and hay nets are mounted directly overhead indoors, and under outdoor shade structures. Swing-arm style feeders are placed around the perimeter of the outdoor exhibits (Fig. 5). These feeders swing outward, allowing keepers to provide the elephants with new feeding opportunities safely while the elephants are in a habitat.

Fig. 5: EA SWING-ARM STYLE FEEDER. ONE OF SEVENTEEN ELEVATED FEEDERS PLACED THROUGHOUT AFRICAN ELEPHANT CROSSING.

Habitat design is only part of the equation. Elephant husbandry staff must manage the habitats in a manner that encourages the elephants to use the habitats' features. Hay and browse are placed in elevated feeding nets and culverts sunk into the ground, and produce is buried. Placing dietary items strategically throughout the exhibit incorporates various feeding strategies and promotes movement throughout the space. Elevated feeders, which are placed at the limit of the elephants' reach, promote exercise that strengthens their necks and shoulders as they reach upward for their food.

In addition to exercise, elevated feeders promote positive social and cognitive activities. In contrast to hay placed on the ground that an elephant can monopolize easily, which can lead to conflict, elevated hay nets can slow the feeding process by requiring the elephant to pull out small amounts of hay at a time. If the elevated nets are placed appropriately, allowing access from multiple sides, elephants cannot monopolize a feeder easily. Placing a feed net just out of reach also affords the elephants the opportunity to use their reasoning and tool use abilities. Elephants will move objects (e.g., logs, balls, stumps) into place and step up onto them in order to reach the food (Fig. 6a). Lastly, these enhancements have resulted in a dramatic increase in cooperative feeding (Fig. 6b) that encourages more natural feeding and social strategies.

FIG. 6. PLACING FOOD ITEMS STRATEGICALLY IN DIFFERENT LOCATIONS AND
CONFIGURATIONS PROMOTES MORE NATURAL FEEDING BEHAVIORS SUCH
AS (**A**) USING TOOLS TO OBTAIN FOOD THAT IS JUST OUT OF REACH MORE
EASILY, AND (**B**) CHOOSING TO FEED SOCIALLY WITHOUT CONFLICT.

Habitat preparation includes managing sand and mud wallows. Sand requires regular maintenance to keep it clean. In addition to daily spot cleaning, keepers turn sand regularly to keep it from becoming compacted and allow wet sand to dry. When indoor sand becomes soiled, it is rotated outdoors so that rain, snow, sun, and periodic turning can clean the sand adequately for rotation back into the building. In addition to cleaning, sand is piled/repiled regularly to provide the elephants with comfortable places to sleep (Fig. 7a). Placing piles in corners and against walls provides sloped areas to lie down while keeping the sand out of heavy traffic areas. Keepers check these piles daily for face and trunk imprints to determine whether the elephants are using them (Fig. 7b).

Fig. 7. Elephants use sand piles located throughout the exhibit for recumbent sleep. In addition to direct observations of recumbence via infrared camera equipment (**A**), face and trunk prints left in the sand indicate that elephants are using the sand piles to lie down (**B**).

The elephants create mud wallows often; however, the keepers maintain them. Because the fitness of wallows can be subjective, keepers assess wallows each day, visually and by smell, to determine whether they require maintenance. Allowing the elephants daily access to the mud wallows not only helps prevent the water from becoming stagnant, but maintains and promotes a valuable feature of the elephants' environment.

MANAGING ELEPHANT BEHAVIOR

CMZ subscribes to a philosophy of continual elephant training throughout the day. Every time a keeper stands in front of an elephant, s/he has the potential to influence that elephant. Therefore, every time a keeper is in front of an elephant, s/he is training that elephant. CMZ uses positive reinforcement training that allows the elephants to choose whether to participate or not. This type of program requires the keepers and elephants to work cooperatively to meet the elephants' health and wellness needs. Husbandry and medical training sessions generally require separating one or more elephants from the rest of the herd. To minimize these separations, the CMZ elephant team focuses on short, but meaningful training sessions, as extended separations can lead to a breakdown of herd dynamics.

Shift training is perhaps one of the most basic behaviors for both keepers and animals. With respect to elephants and herd dynamics, shifting is a critical process. Elephants are stationed while hydraulic doors are operated, and move from one area to another under a keeper's direction, who often follows along with them while the elephants are shifting. At these times, keepers need to work with the elephant's social hierarchy rather than against it. This often means shifting the elephants in the order in which they choose. This is not always possible, particularly when the keepers need to separate one elephant from the herd for husbandry or medical purposes. In these instances, keepers must use care not to put an elephant in a situation that will make them uncomfortable or worse yet, a target for a more dominant animal. To allow safe shifting according to social dynamics, elephants are shifted using a flexible degree of control. Shifting under "flexible control" means keepers allow the elephant to demonstrate natural postures and express normal body language while shifting. For example, a submissive animal that is being stationed may be allowed to drift their hind end toward a dominant animal as a sign of submission. Allowing the elephants to display their natural behaviors while shifting promotes herd cohesiveness.

SCIENCE-BASED MANAGEMENT AND EVALUATION

Complementing the direct care provided to the elephants, research staff members help collect, analyze, and disseminate data about elephant behavior, physiology, and exhibit use. Having access to scientific capacity on site afforded us the opportunity to undertake an extensive Post-Occupancy Evaluation (POE) of AEC when it opened in 2011. POE is used to determine the effects of a designed environment on its occupants (Maple & Finlay, 1987). In our case, we were interested in evaluating the effects of our new, larger, naturalistic exhibit on the elephants themselves. However, the opening of AEC did not simply change the elephants' space. We also changed many of our husbandry practices and general elephant management philosophy at that time, which resulted in a learning curve for both the elephant care team and the elephants. We also introduced three new elephants from three different institutions to the three who had been housed together previously in the original exhibit, which created a new, more complex social dynamic. The fact that all of these major changes occurred—and that they occurred at the same time—made a thorough POE necessary so that we could determine which aspects of our new program were working for the elephants, and which could be improved.

In 2005, CMZ began collecting behavior data on the elephants as part of a study that investigated seasonal variation in behavior in our northern climate zoo (Mueller Dennis, Willis, Simone, & Lukas, 2013). In the last decade, several additional research projects have resulted in the accumulation of a longitudinal dataset on the six elephants we housed in Cleveland. However, behavior changes are only part of the story. It is widely accepted that studies incorporating multiple measures of welfare from across disciplines are most effective in identifying factors associated with compromised welfare (Whitham & Wielebnowski, 2013). In this POE, we focused on measures indicative of health and fitness, as research has suggested that obesity is associated with a host of chronic health conditions in captive elephants (summarized in Morfeld et al., 2016). Banked serum samples were used to measure leptin (a measure of body fat: see Considine et al., 1996; Morfeld & Brown, 2014), glucose:insulin ratio (an indicator of insulin sensitivity: see Divers, 2008), and progesterone (a measure of ovarian cyclicity: see Brown et al., 2016 for a discussion of acyclicity in elephants). This extensive database allows us to determine the way in which the health and behavior of individual elephants has changed over time in response to housing and management changes.

There is an increasing recognition that animal welfare and quality of life are best measured from the perspective of the individual. An individual's unique experiences, preferences, temperament, and life history influence his/her perceptions of the environment and overall well-being (Whitham & Wielebnowski, 2013). Mason and Veasey (2010) argued that elephant husbandry should not only be evidence-based, but should consider each elephant's own perspective. Our long-term dataset has the two-fold benefit of not only allowing us to determine the way in which management changes affect the herd, but each elephant as well.

In the first phase of our POE, we compared behavior data collected in the original exhibit with those collected in the first year after the move into AEC (2012). One might expect that having nearly four times more indoor and outdoor space would naturally increase our elephants' degree of activity. However, the data did not support this prediction. Although time spent feeding and foraging increased more than 10% in the new exhibit, locomotion and time spent inactive did not change. The elephants were no more active in the new exhibit than in the previous one. With respect to health, the two newest females were gaining weight, showing increasing leptin (body fat) values, and decreasing glucose:insulin values (indicating increased potential for insulin resistance). However, the two original females were maintaining their body weight, showing decreases in leptin values, and increases in glucose:insulin values. Therefore, although the entire herd showed no changes in activity levels overall between the old exhibit and the new one, there seemed to be individual variation in the way in which each was responding physiologically to this change.

The exhibit itself was not the only variable we considered during this first year. We were also working to integrate the original three females (45, 36, and 31 years old), two new females (30 and 31 years old), and a male (33 years old) into one cohesive, yet completely unrelated, herd. In the midst of this first year transition, our oldest female passed away at the age of 46, thereby altering the group's social dynamics further. During this process of integration, the animal care team was managing multiple small social groups in different spaces, which limited each individual's choice of space, a factor that helped explain the lack of change in activity levels during this time. To promote cohesiveness in the group and minimize food competition, the animal care team also provided extra hay for all elephants during introductions to prevent any one individual from monopolizing the food supply. The two newest females reportedly had a history of food aggression and possessiveness, so the team was willing to accept some weight gain during this transition to promote herd cohesion.

Over the next year, the social dynamics began to stabilize and we were able to manage the females as one herd, with the male joining them for increasing amounts of time. This allowed the entire exhibit space to be used regularly. However, simply providing space is not always enough, and we wanted to encourage the elephants to use it fully. Now that we had a relatively cohesive herd, we could begin to address weight gain and health through long-term changes in feeding and housing strategies. We began installing elevated feeders throughout the indoor space to encourage the elephants to engage more muscle groups and work a little harder for their food. All of these elephants had spent most of the last 30+ years feeding from piles of hay on the ground. This seemingly small and simple addition to the exhibit offered a brand new physical and mental challenge. In addition to the elevated feeders indoors, we began to offer more browse and shifted the herd through different spaces several times a day to encourage exploration and exercise.

In the fall of 2013, we began the second phase of the POE to determine how these new changes were affecting our elephants. Behavior data collected between 2013 and 2014 revealed a significant increase in activity levels across the entire herd. Locomotion increased by nearly 5% and time spent inactive decreased by half. Physiologically, body weight was leveling off (no longer increasing), leptin was beginning to decrease, and glucose:insulin was increasing in all four females. In addition, progesterone analysis showed that not only were all four females now cycling regularly, but they also were cycling synchronously, further demonstrating social cohesion in this group.

During this period, we also began to use infrared cameras to monitor nighttime behavior. Since they moved into AEC, all females have been given free access to multiple spaces and to each other for the 14 hours that animal care staff is not present each day. When monitoring began, recumbent sleep time varied between individuals, some sleeping 5-7 hours per night, and some never lying down. Over time, the females not only began to lie down nightly, but they chose to lie together more often. One of the most encouraging behaviors we observed during this video monitoring was the increase in social play among these four unrelated adult females. Chasing, trunk wrestling, and rolling together in the sand were common occurrences after animal care staff left for the day. All of these data together indicated that we were moving in the right direction for elephant wellness at CMZ.

Nonetheless, we are not finished. We believe strongly that there is never an "end goal" in animal care and wellness. There is always room for improvement, and just because we are seeing positive results now does not mean we can be complacent. Since 2014, we have continued to improve our elephant program and habitats. We installed additional elevated feeders in the outdoor yards, and are continuing to expand our browse program and monitor caloric intake carefully. The male spends several hours with the females daily, and we strive continually to create environments that encourage exercise and exploration. Though he has been determined to be sterile, the male has bred each of the females. Thus, the herd is able to experience this behavior without the danger of a pregnancy in these nulliparous, aging females.

We compared the behavior data collected in 2015 to those from previous years and continue to see some positive trends. Inactivity now accounts for less than 3% of the daytime activity budget, and investigative behaviors have increased significantly. However, locomotion has now decreased significantly, which initially was surprising. However, time spent feeding socially has more than doubled, so one explanation could be that the elephants are choosing to stand together and feed in one location rather than travel to different areas to feed separately. Whether this result can be interpreted as good or bad is certainly debatable. What we can conclude is that this information confirms our belief that nothing in animal behavior and welfare is static, and we should devise new strategies for improvement continually.

Our elephant program has experienced many changes since AEC opened in 2011. Through evaluation, we learned that having a larger, more complex exhibit does not produce immediate improvements in behavior and health. Social dynamics and opportunities to express natural behaviors have as much effect on wellness as does the physical space. Monitoring both behavior and health can provide a more complete picture of what is happening with the animals in our care, and allows any potential issues to be addressed sooner rather than later. Continuing to improve, evaluate, and adjust over time is key in promoting optimal health and welfare, regardless of the species. Integration is the core of CMZ's management philosophy for the elephants and all species in our care. Having a strong partnership and reciprocal relationship among research, animal care, and management is critical to ensure wellness, and we strive to reach that goal every day.

ACKNOWLEDGEMENTS

The authors would like to thank the entire Elephant Care team at Cleveland Metroparks Zoo, including our elephant consultant, Alan Roocroft, of Elephant Business, Inc., keepers, curators, Veterinary staff, and Conservation & Science staff for their dedication to elephant care and wellness. In addition, the authors would like to thank Travis Vineyard, Andi Kornak, Albert Lewandowski, DVM, and Kristen Lukas, Ph.D. for their helpful edits and suggestions for this manuscript, and Chris Kuhar, Ph.D. for his leadership in our evolving zoological philosophy.

REFERENCES

Brown, J. L., Paris, S., Prado-Oviedo, N. A., Meehan, C. L., Hogan, J. N., Morfeld, K. A., & Carlstead, K. (2016). Reproductive health assessment of female elephants in North American zoos and association of husbandry practices with reproductive dysfunction in African elephants (*Loxodonta africana*). *PloS ONE, 11*(7).

Carlstead, K., Brown, J. L., & Seidensticker, J. (1993). Behavioral and adrenocortical responses to environmental changes in leopard cats (*Felis bengalensis*). *Zoo Biology, 12*(4), 321-331.

Considine, R. V., Sinha, M. K., Heiman, M. L., Kriauciunas, A., Stephens, T. W., Nyce, M. R., et al. (1996). Serum immunoreactive-leptin concentrations in normal-weight and obese humans. *New England Journal of Medicine, 334*(5), 292-295.

Divers, T. J. (2008). Endocrine testing in horses: Metabolic syndrome and Cushing's disease. *Journal of Equine Veterinary Science, 28*(5), 315-316.

Kuhar, C. W., Kornak, A. M., & Lukas, K. E. (2015). Beyond animal welfare science. *WAZA Magazine, 16*, 30-33.

Maple, T. L., & Finlay, T. W. (1987). Post-occupancy evaluation in the zoo. *Applied Animal Behaviour Science, 18*(1), 5-18.

Mason, G. J., & Veasey, J. S. (2010). How should the psychological well-being of zoo elephants be objectively investigated? *Zoo Biology, 29*(2), 237-255.

Melfi, V. A. (2009). There are big gaps in our knowledge, and thus approach, to zoo animal welfare: A case for evidence-based zoo animal management. *Zoo Biology, 28*(6), 574-588.

Morfeld, K. A., & Brown, J. L. (2016). Ovarian acyclicity in zoo African elephants (*Loxodonta africana*) is associated with high body condition scores and elevated serum insulin and leptin. *Reproduction, Fertility and Development, 28*(5), 640-647.

Morfeld, K. A., Meehan, C. L., Hogan, J. N., & Brown, J. L. (2016). Assessment of body condition in African (*Loxodonta africana*) and Asian (*Elephas maximus*) elephants in North American zoos and management practices associated with high body condition scores. *PloS ONE, 11*(7).

Mueller, J. E., Dennis, P. M., Willis, M. A., Simone, E. A., & Lukas, K. E. (2013). Seasonal and diurnal variations in African elephant (*Loxodonta africana*) and Black rhinoceros (*Diceros bicornis*) behavior in a Northern climate zoo. *International Journal of Comparative Psychology, 26*(4), 304-323.

Shepherdson, D. J., Carlstead, K. C., & Wielebnowski, N. (2004). Cross-institutional assessment of stress responses in zoo animals using longitudinal monitoring of faecal corticoids and behaviour. *Animal Welfare-Potters Bar Then Wheathampstead, 13*, S105-S114.

Stoinski, T. S., Lukas, K. E., & Maple, T. L. (1998). A survey of research in North American zoos and aquariums. *Zoo Biology, 17*(3), 167-180.

Whitham, J. C., & Wielebnowski, N. (2013). New directions for zoo animal welfare science. *Applied Animal Behaviour Science, 147*(3), 247-260.

Chapter 6

A HOLISTIC APPROACH TO ELEPHANT WELLNESS

Why Reinforcement, Choice, and Employee Retention Matter

Megan L. Wilson

Georgia State University, Atlanta, Georgia

AUTHOR'S NOTE

The data presented in this paper were previously published in *Zoo Biology* in 2015 (34: 431–437). Correspondence concerning this chapter should be addressed to Megan Wilson, Department of Psychology, 1189 Urban Life Building, Georgia State University, Atlanta, GA, 30313, mwilson72@gsu.edu.

INTRODUCTION

This paper describes previous published research (Wilson, Perdue, Bloomsmith, & Maple, 2015) conducted to provide evidence to support the following questions: 1) to what degree are positive and negative reinforcement used in both free contact and protected contact management systems, and 2) do these different management systems differentially affect elephant welfare? In the study, we used all-occurrence sampling (Altmann, 1974) to collect data on the interactions between elephant care professionals (ECPs) and elephants during regularly scheduled bathing of the elephants; the result was a running tally of all ECP and elephant behaviors. Due mostly to the small sample size, we examined the data using descriptive statistics. The present paper reviews these findings and stresses the importance of adopting a holistic approach to elephant management, which emphasizes appropriate use of reinforcement, provision of choice, and employee retention, all with the goal of maximizing wellness.

COMPARING FREE
AND PROTECTED CONTACT

Concerns for personnel safety and, in some cases, animal welfare, have caused many zoos to change the way elephants in human care are managed. Once the mainstay for elephant management, the "free contact" system for elephant management was challenged by the success of an alternative type of management, deemed "protected contact" in the early 1990s (see Desmond & Laule, 1991; Whittaker & Laule, 2009). Proponents of both management systems argue that their respective system holds the advantage; however, distinct differences between the two prompt questions that have yet to be answered, particularly those questions related to elephant wellness. For example, the use of different types reinforcement is often cited as a clear difference between the systems. Although ECPs in both systems use positive reinforcement, protected contact should rely on this type of reinforcement to encourage cooperation on the part of the animal (Desmond & Laule, 1994). In addition, negative reinforcement is not utilized unless all positive options have been exhausted (Whittaker & Laule, 2009), whereas ECPs use an ankus (or bullhook) as negative reinforcement (e.g., pressure is applied behind the elephant's foot with the ankus, and the elephant moves her foot to avoid the pressure) in free contact. This leads to two primary questions that we attempted to answer: 1) to what degree are positive and negative reinforcement used in both free contact and protected contact management systems, and 2) do these different management systems differentially affect elephant wellness?

A number of years ago, when the elephants at Zoo Atlanta made the transition from free to protected contact, we set out to answer these questions (Wilson et al., 2015). In brief, the elephants were bathed on a regular basis in the elephant program, and thus, we used the interactions between elephant care professionals (ECPs) and elephants to document the behavior of both ECPS and elephants. We used all-occurrence sampling (Altmann, 1974), which gave us a running tally of all trainer and elephant behaviors that occurred during a bath. Due mostly to the small sample size (three elephants), we examined the data using descriptive statistics. The present paper will outline the primary findings and interpret them in relation to reinforcement, choice, and employee retention.

KEY FINDINGS

Although there were some similarities between the baths given in both management systems (e.g., mean bath length, ECP behavior), clear differences emerged regarding the use of reinforcement when we examined the data (summarized below):

- The ECPs use of the ankus/hook (negative reinforcement) occurred at a much lower rate in free contact than the use of positive reinforcement (which included food, verbal, and touch combined).

- The ECPs use of the ankus/hook (negative reinforcement) occurred at a much lower rate in free contact than positive reinforcement in protected contact.

- The use of the ankus/hook by the trainers in free contact occurred at almost the same rate as food reinforcement.

- Positive reinforcement in the form of food and verbal praise was delivered to the elephants at much higher rates in protected contact.

- We also documented clear differences between free contact and protected contact in relation to elephant compliance (summarized below):

- Latencies between verbal commands and the elephants' behaviors were generally longer, and the mean percent of "refusals" was generally higher for all elephants in protected contact.

Taken together, we concluded that the ECPs did not rely on positive reinforcement during the observed baths in free contact to elicit desired behaviors from the elephants, but instead used nearly equivalent rates of positive and negative reinforcement. It would appear, then, that free contact is not the most positive, least intrusive way to manage elephants. Additionally, measures of latencies and compliance suggest that the elephants may have begun to exercise choice during the study period and that managing this herd of elephants in PC allowed them access to additional opportunities for positive reinforcement, thus potentially improving their welfare.

A HOLISTIC APPROACH
TO ELEPHANT MANAGEMENT

In order to begin to answer whether the two systems differentially affect wellness, it's important to look at the results in the context of several factors that are believed to positively affect animal welfare: positive reinforcement, choice/control, and relationships between animals and their caregivers.

Positive Reinforcement. Positive reinforcers in the form of food were offered to the elephants almost eight times more frequently in protected contact, although positive reinforcement was used in both systems. Positive reinforcement training has been linked to reduced fear in animals; conversely, negative outcomes are associated with management systems that incorporate negative reinforcement and punishment (see Maple, Bloomsmith, & Martin, 2009, for a discussion of this topic). In the studied herd of elephants, they had access to more opportunities for positive reinforcement, thus potentially improving their wellness. It follows that an added benefit to this reliance on positive reinforcement would be a more positive elephant-ECP relationship (see Ward & Melfi, 2013, for a discussion about positive reinforcement and its effect on zoo animal training) and that this relationship may persist over time, as has been shown in equids (Sankey, Richard-Yris, Leroy, Henry, & Hausberger, 2010; Sankey et al., 2010).

Choice/Control. Latencies between verbal commands and the elephants' behaviors were generally longer in protected contact, and the mean percent of "refusals" was generally higher. These results suggest that the elephants were choosing when to comply with certain requests made by the ECPs in the new management system. In other words, the elephants were likely exercising some level of control over their environment that wasn't available to them in free contact. Although the results of this study are limited by the small sample size, there is growing support that choice and control are beneficial for animals in captivity. Increases in available control may be related to decreases in undesirable behaviors (i.e., stereotypic behavior: Bloomsmith, Keeling, & Lambeth, 1990; Bloomsmith & Lambeth, 2000; Bloomsmith, Baker, Lambeth, Ross, & Schapiro, 2000) and reduced physiological manifestations of stress (Hanson, Larson, & Snowdon, 1976), as well as promoting desirable behaviors such as greater exploration (Mineka & Hendersen, 1985; Innis & McBride, 2008) and improved cognitive performance (Washburn, Hopkins, & Rumbaugh, 1991). When given the "opportunity to choose," such as the order in which computer tasks are completed, monkeys will do so (Perdue, Evans, Washburn, Rumbaugh, & Beran, 2014). Taken as a whole, the importance of choice has probably been underestimated as it relates to elephant management and specifically as it relates to training practices.

Elephant managers should take choice and control one step further and promote autonomy in elephants. Autonomy, a term borrowed from the literature on human wellness, is best thought of as independence or freedom, where the individual is not controlled by an external entity. Giving elephants the ability to make choices and have some degree of control over their environment is best supported by protected contact management. Providing elephants with autonomy may be critical to achieving wellness and perhaps necessary for elephants to thrive in human care.

Employee Retention. Our data indicate that the animal care professionals made positive physical contact, in the form of patting and rubbing, at lower rates in protected contact. It may be that this is the result of perception on the ECPs' part, in that they saw protected contact management as a "hands-off" approach, and that barriers limited their ability to touch the elephants. Although a possibility, it seems that ECP attitudes toward protected contact are the more likely cause. In our 2015 study, eight trainers were observed during the course of the study, but only three of them were observed in both free and protected contact.

Some animal care professionals are reluctant to change from free to protected contact because they worry that their relationships with the elephants will be negatively impacted. Specifically, they express concern that, in protected contact, they will lose both the special relationships that they have developed with the elephants (Desmond & Laule, 1993) and the opportunity to move freely among the elephants as a herd member (Priest, 1992a). It is possible that these concerns, along the questions about providing proper care to elephants in a protected contact system (Priest, 1992b), may result in an ECP leaving a particular zoo when its elephant program transitions to protected contact. These concerns do not appear to be isolated to the ECPs observed in our study. Managers need to spend as much time working with the ECPs during the transition to protected contact as they do with the elephants to assess and improve job satisfaction, as well as create consistency in the program.

CONCLUSIONS AND FUTURE STUDY

There are many directions for future study that will help us achieve a holistic approach to elephant management, which maximizes elephant wellness. First, when possible, managers should evaluate the training program in both systems to fully understand the effect of the transition on elephant behavior. Second, observations in our 2015 study occurred during the first eight months of the transition, and trainers were often providing food to the elephants at very high rates. It is likely that this has changed over the years, and currently, we are working to replicate the study to assess how the rate of reinforcement in protected contact has changed since the transition. Additional institutions should collect data on the rate at which reinforcement is provided in their current management system. Third, managers should document trainer attitudes regarding the transition to protected contact to increase job satisfaction and decrease employee turnover, which will provide additional consistency to the program.

The literature tells us that animals are positively affected when they are managed in ways that rely on positive reinforcement, allow them choice and control over their environment, and aim for a positive relationship between animal and trainer. When thinking about elephant wellness, we can to assess each system based on these factors. On the surface, it appears as if a traditional free contact program doesn't incorporate much of what was just described but that protected contact does. We've seen how we can positively affect animal wellness when we strive to manage other species with this holistic approach. The question we need to ask ourselves is whether elephants can benefit from this this approach in the same way that other species do. The answer seems clear: they will.

REFERENCES

Altmann, J. (1974). Observational study of behavior: Sampling methods. *Behavior, 48,* 227–265.

Bloomsmith, M. A., Baker, K. C., Lambeth, S. P., Ross, S. K., & Schapiro, S. J. (2000). Is giving chimpanzees control over environmental enrichment a good idea? In *The apes: Challenges for the 21st century, conference proceedings* (pp. 88–89). Brookfield, IL: Chicago Zoological Society.

Bloomsmith, M. A., Keeling, M. E., & Lambeth, S. P. (1990). Videotapes: Environmental enrichment for singly housed chimpanzees. *Lab Animal, 19,* 42–46.

Bloomsmith, M. A., & Lambeth, S. P. (2000). Videotapes as enrichment for captive chimpanzees (*Pan troglodytes*). *Zoo Biology, 19,* 541–551.

Desmond T., & Laule G. (1991). Protected-contact elephant training. In *AZA Proceedings.* Wheeling, WV: Association of Zoos and Aquariums.

Desmond T., & Laule, G. (1993). *The politics of protected contact.* In *AZA Proceedings, Omaha, NE.*

Desmond T., & Laule G. (1994). Use of positive reinforcement training in the management of species for reproduction. *Zoo Biology, 13,* 471–477.

Hanson, J. D., Larson, M. E., & Snowdon, C. T. (1976). The effects of control over high intensity noise on plasma cortisol levels in rhesus monkeys. *Behavioral Biology, 16,* 333–340.

Innis, L., & McBride, S. (2008). Negative versus positive reinforcement: An evaluation of training strategies for rehabilitated horses. *Applied Animal Behavior Science, 112,* 357–368.

Maple, T., Bloomsmith, M. A., & Martin, A. L. (2009). Primates and pachyderms: A primate model of zoo elephant welfare. In D. L. Forthman, L. F. Kane, D. Hancocks, & P. F. Waldau (Eds.), *An elephant in the room: The science and well-being of elephants in captivity* (pp. 129–153). North Grafton, MA: Tufts University Press.

Mineka, S., & Hendersen, R. W. (1985). Controllability and predictability in acquired motivation. *Annual Review of Psychology, 36,* 495–529.

Perdue, B., Evans, T., Washburn, D., Rumbaugh, D., & Beran, M. (2014). Do monkeys choose to choose? *Learning and Behavior, 42,* 164–175.

Priest, G. (1992a). *Elephant training: Creating contrast between two training systems.* In *AAZPA Proceedings, Tucson, AZ.*

Priest, G. (1992b). *The development of protected contact elephant management.* In *AAZK Proceedings, San Diego, CA.*

Sankey C., Richard-Yris, M. A., Leroy, H., Henry, S., & Hausberger, M. (2010). Positive interactions lead to lasting positive memories in horses (*Equus caballus*). *Animal Behavior, 79,* 869–875.

Sankey, C., Richard-Yris, M.A., Henry, S., Fureix, C., Nassur, F., & Hausberger, M. (2010). Reinforcement as a mediator of the perception of humans by horses (*Equus caballus*). *Animal Cognition, 13,* 753–764.

Ward, S. J., & Melfi, V. (2013). The implications of husbandry training on zoo animal response rates. *Applied Animal Behavior Science, 147,* 179–185.

Washburn, D. A., Hopkins, W. D., & Rumbaugh, D. M. (1991). Perceived control in rhesus monkeys (*Macaca mulatta*): Enhanced video-task performance. *Journal of Experimental Psychology: Animal Behavior Processes, 17,* 123–129.

Whittaker, M., & Laule, G. (2009). Protected contact and elephant welfare. In D. L. Forthman, L. F. Kane, D. Hancocks, & P. F. Waldau (Eds.), *An elephant in the room: The science and well-being of elephants in captivity* (pp. 181–188). North Grafton, MA: Tufts University Press.

Wilson, M. L., Perdue, B. M., Bloomsmith, M. A., & Maple, T. L. (2015). Rates of reinforcement and measures of compliance in free and protected contact management systems. *Zoo Biology, 34,* 431–437.

Chapter 7

COGNITIVE RESEARCH
AND ELEPHANT WELLNESS

Lauren Highfill, Eckerd College; Otto Fad, Precision Behavior; and
Jessica Spencer, Busch Gardens Tampa Bay

AUTHORS' NOTE

We would like to dedicate our chapter to the memory of Dr. Stan Kuczaj, who was our inspiration and mentor. Stan left this world too early but left behind an impressive legacy of accomplished students, devoted collaborators, fortunate animals, and grateful friends, in addition to his prodigious contributions in the areas of comparative psychology and animal cognition. He will be missed but always remembered fondly and with sincere admiration and heartfelt gratitude.

INTRODUCTION

Scientific research and the modern zoo have had a long relationship. Historically however, zoos have focused their research efforts primarily on veterinary medicine, conservation, and animal welfare, while research on the cognitive abilities of animals, such as problem-solving, memory, and visual discrimination, has focused primarily on laboratory animals (e.g., rodents, birds, and primates: Plotnik, de Waal, Moore, & Reiss, 2010). However, in recent years, there has been increasing interest in exploring animal cognition with zoo animals through noninvasive psychological testing paradigms. This provides the field of animal cognition with access to animals that are not kept easily in a laboratory setting. For example, Terry Maple and his colleagues at Zoo Atlanta investigated whether gorillas could judge quantity (Anderson et al., 2005), and Stan Kuczaj and colleagues tested whether bottlenose dolphins could use a symbolic keyboard at Epcot's Living Seas (Xitco, Gory, & Kuczaj, 1999). In this chapter, we discuss the advantages of conducting cognitive research with zoo animals, specifically elephants.

BENEFITS OF COGNITIVE RESEARCH IN ZOOS

First, cognitive research provides an opportunity to engage visitors with science, watch research as it happens, and foster a better understanding of the animals themselves. An excellent example of this is the Living Links Research Centre at the Edinburgh Zoo. Opened in 2008, Living Links is a field station and research center of the University of St. Andrews, established in partnership with the Royal Zoological Society of Scotland and Edinburgh Zoo. It has large indoor and outdoor enclosures in which capuchin monkeys and squirrel monkeys live together. Living Links was designed specifically to support studies by scientists from neighboring universities. The research center encourages the public actively to observe primate behavioral and cognitive research and read about past studies described in the signage posted around the exhibit. A recent study examined the effect of live science demonstrations on guest behavior at Living Links and found that dwell time within the center overall increased significantly when research was occurring (Bowler, Buchanan-Smith, & Whiten, 2012). The concept is that when they stay longer, guests potentially learn more about the animals, which increases their interest overall and perhaps even motivates them to contribute to conservation efforts after seeing animals in a new way.

Cognitive research in the zoo environment is beneficial from a conservation perspective in that findings not only add to the body of scientific knowledge about species but can complement or examine in finer detail the discoveries made in observational studies. Much of the applied work required to study cognition would be difficult or impossible to undertake in the field. Further, from an ethical standpoint, it would be inadvisable to do so because manipulating aspects of the environment intentionally influences the behavior of wild animals.

Elephant populations in Asia and Africa are threatened by human-elephant conflict attributable to habitat encroachment and fragmentation and are being displaced by rapidly expanding human populations. Asia (4.4 billion) and Africa (1.2 billion) are currently first and second, respectively, in population, with continued growth predicted in the coming decades. Showing elephants engaged in cooperative tasks with their caretakers may help observers envision a cooperative relationship between elephants and humans as they work together in close proximity rather than function in isolation from each other, or worse, compete for resources.

From the animals' perspective, participating in cognitive research can serve as environmental enrichment that increases the psychological well-being of the study species. Cognitive research allows the animals to think and control aspects of their environment. We know from the data on contra-freeloading that when offered a choice between free food and that which requires some effort, most animals prefer to work for food (Neuringer, 1969). A zoo animal that must work for its reward mimics what they would face in the wild. Many of the cognitive research paradigms use food as a reward, so this is one way to engage the animals in mental stimulation to receive their reward. Of course, cognitive research can be challenging for animals, as they do not always get the right answer. However, neither do their wild counterparts. Some theorists believe that some level of frustration (which ultimately turns to success) is good for animals that live in environments controlled by humans (Meehan & Mench, 2007).

Cognitive research also can benefit the animal caretakers, as it can expand their relationship with the animals. It affords them unique opportunities to observe the way in which individuals react to different situations, which provides a novel way to learn more about their animals. Further, from both the researcher's and animal caretaker's perspectives, cognitive research allows us to acquire a better understanding of the animals. Consider this quote: "Working to improve captive animal care without an understanding of how animal minds work, or the scope in which they perceive and interact with their social and physical environments, is akin to drawing a map without knowledge of any landmarks or bearings" (Ross, 2010).

This quote highlights the connection between our understanding of an animal's cognitive abilities and our ability to provide them with the best animal care. It is established well that wild animals face many challenges that require the use of evolved cognitive skills (e.g., remembering locations of watering holes and food sources, avoiding predators). In contrast, animals in zoos tend to live in highly predictable and controlled environments and may have insufficient opportunities to exercise their cognitive skills. Fortunately, providing zoo animals with more cognitively challenging enrichment programs has increased in recent decades. However, to maximize animal wellness, individuals should possess sufficient skill to master the challenges and be motivated to do so. Thus, the more we understand about the way animals' minds work, the better we can care for them.

Finally, zoos themselves also may benefit from cognitive research undertaken with their animals. Building a robust science to support wellness will inform care standards and may help counteract anti-zoo narratives. Zoo-goers who observe engaged animals solving problems may acquire an entirely different perspective than that derived from watching animals engaged in less complex behavior.

ELEPHANT COGNITION

Anyone who has ever worked with or observed elephants would agree that they are intelligent animals. Their brains are large and complex, and they have cognitive abilities similar to those observed in humans and nonhuman primates (Shoshani, Kupsky, & Marchant, 2006). Accordingly, elephants have demonstrated advanced cognitive abilities both in the wild and under human care. For example, elephants show remarkable ability to use tools and solve problems. Both in the wild and in zoos, they use branches to swat flies or scratch themselves (Chevalier-Skolnikoff & Liska, 1993). Wild elephants even have been observed to drop boulders onto electric fences to turn off the electricity and leave a fenced area (Poole, 1996). Elephants under human care also have been tested on a number of cognitive tasks similar to those that have been administered to primates and dolphins, e.g., number discrimination (Perdue, Talbot, Stone, & Beran, 2012) and self-recognition (Plotnik, de Waal, & Reiss, 2006). In a recent study, Plotnik, Lair, Suphachoksahakun, & de Waal (2011) demonstrated that elephants could learn to coordinate their actions with a partner in a task that required two individuals to pull two ends of the same rope simultaneously to obtain a reward. In addition to pulling the rope cooperatively, the elephants inhibited their pulling for up to 45 seconds if a partner's arrival was delayed. They also seemed to understand that there was no point in pulling if the partner could not access the rope.

Most of what is known about elephant cognition has come indirectly from studying wild populations or working elephants in Asian tourism or logging industries. However, zoo-based research has been conducted successfully, and several topics have been explored, including long-term memory, theory of mind, social learning, causal reasoning, and numerical competency. One of the first published studies to examine the cognitive abilities of a zoo elephant was conducted at Münster Zoo on a single Asian elephant (Rensch, 1956). For this study, the subject was presented boxes with lids that had been painted with distinctive patterns. The elephant was trained successfully to discriminate between 20 pairs of patterns to receive a food reward. The same elephant was tested again after a year break and demonstrated a remarkable retention for the task.

Another study at the Detroit Zoological Institute tested two Asian elephants on their ability to demonstrate theory of mind by examining their "begging" behavior from two possible human experimenters (Nissani, 2004). In each condition, only one of two experimenters' faces were visible to the subject. For example, in one condition, one of the experimenters wore a bucket over her head while the other experimenter set a bucket on her shoulder next to her face. Overall, the results indicated that elephants were able to beg from the correct experimenter, thus suggesting elephants possess an understanding of human gaze. During follow-up trials, Nissani (2004) had a single keeper stand either facing toward the subject or facing away from the subject. Overall, the elephant would beg silently when the keeper was facing her. However, she would forcefully blow air on the keeper when the keeper was facing away from her. Overall, the results seemed to suggest that elephants may outperform chimpanzees on such tasks.

To further explore theory of mind capabilities of elephants, three elephants were tested on their ability to use a mirror at the Bronx Zoo (Plotnik et al., 2010. Overall, when presented with a large mirror, the elephants performed similarly to apes and dolphins in the same type of task. Ultimately, the elephants demonstrated an understanding of the mirror by using it to explore the inside of their mouths and even investigating a mark on their face.

More recently, researchers at the San Diego Zoo examined the possibility of social learning within a herd of six female elephants (Greco, Brown, Andrews, Swaisgood, & Caine, 2013). Novel puzzle feeders which could be solved in two distinct ways were used. Overall, evidence supported that observing a conspecific completing a novel task increased the subject's interest in the task but did not predict which of the two ways the subject would solve the puzzle feeder. At the Smithsonian National Zoo in Washington, DC, Foerder, Galloway, Barthel, Moore, & Reiss (2011) witnessed spontaneous problem-solving when one of the subjects moved a large cube in order to stand on it to reach food that had been placed out of reach. Finally, Perdue et al. (2012) tested two African elephants at Zoo Atlanta on their ability to judge quantities of food. They found that elephants performed similarly to primates and other animals when choosing between different amounts of food. Overall, it seems that cognitive research can be conducted successfully at zoological institutions.

ELEPHANT COGNITION AT BUSCH GARDENS TAMPA BAY

While the evidence supports the notion that elephants possess high levels of intelligence, there is still much that we do not know and many more questions that need to be tested empirically with this species. In doing so, zoos can enhance the welfare of their herds by applying the same elephant management philosophy and practices to the design of a cognitive study, including a commitment to positive reinforcement training, providing a choice-based environment, and using variability. This is not to say that any study qualifies automatically as enriching for the participants, but that specific design elements can contribute to enhancing wellness while still maintaining sound, scientific rigor.

If we are to consider participating in behavioral research an avenue for enrichment, we must analyze it according to the features that define enrichment. The research session should create change in the environment, provide behavioral choices, elicit species-specific behaviors, and stimulate mental and/or physical exercise. Ultimately, it should enhance animal welfare. At Busch Gardens Tampa Bay, the resident elephants participated in a cognitive study we will use as an example to explore each of these facets (Highfill, Spencer, Fad, & Arnold, 2016).

The Asian elephant (*Elephas maximus*) herd at Busch Gardens Tampa Bay consists of one male and five female adults. The herd has been managed since 2004 in a protected-contact setting in which trainers use only positive reinforcement training (PRT) in a choice-based environment. The program touts a dynamic behavioral management style in which changes in the elephants' physical and social environments are made throughout the day, as well as a commitment to variability. Overall, the elephants' environment can be described as an open economy. Food, water, and social interactions with conspecifics are always available in an enriched environment with many stimuli that may compete with training sessions or research trials. Elephants initiate interactions with trainers frequently and can choose not to participate in sessions without negative consequences. In fact, trainers do not use the word "no," nor does any analog for it exist in their training "vocabulary." Because there are no corrections, trainers must recognize and reinforce desirable (and in some cases merely acceptable) behavior.

Research sessions are conducted outdoors in the main habitat at the Elephant Interaction Area, where members of the herd join trainers frequently for interaction in the form of learning, husbandry, and relationship sessions through a husbandry gate. This area was constructed to enable guests to view these interactions and procedures.

The study was a systematic replication of the Irie-Sugimoto, Kobayashi, Sato, and Hasegawa (2008) experiment, in which two elephants were tested in a free-contact setting. The nature of the study involved a means-end task in which elephants could demonstrate whether they understood the means necessary to achieve a goal, in this case, accessing distant food. To test this, each elephant had the opportunity to choose one of two trays, in which only the near end of the trays was within reach. The far end of one of the trays always was baited with a highly valued food item, so that the elephant had to pull the tray toward him/her to access the food. The other tray either was not baited (Condition A), baited on the floor next to the tray (Condition B), or baited on the tray but with a gap in the middle of the tray (Condition C) so that none of the options was viable to obtain the food.

The results indicated that all six elephants succeeded in Condition A in 30 trials and Condition B in 40 trials or less. None of the elephants performed above chance levels on Condition C within 50 trials. In contrast, at least one of the elephants in the original Irie-Sugimoto et al. (2008) study succeeded in all three conditions but learned over time and took up to 250 trials to demonstrate proficiency. The Busch Gardens elephants appeared to show a spontaneous understanding of the means-end relationship, at least in the first two conditions. The final condition was more difficult, as it involved the relatively more complex concept of connectedness. It could also be argued that this condition deviated to a greater degree from the experiences the elephants may have had in their environment (e.g., pulling on a tree branch to access the leaves), among numerous other possible explanations.

Regardless of the results obtained, each elephant's keen willingness to engage actively in the task supports the notion of its enrichment value, but we must parse out the relevant factors, beginning with the change in the environment. Asking each elephant to approach the Elephant Interaction Area was not novel, but the physical set-up in preparation for the trials was, including the specific configuration of open windows, an additional wooden attachment to the husbandry wall, and the view of the supplies. At the beginning of testing, the trays themselves were also a novelty to be explored. Moreover, during the time of day that the elephants normally would be involved in other activities, their active participation in the interactive research session itself constituted a notable change in their environment.

Because each elephant was tested individually, they had an opportunity for more one-on-one interaction with trainers without having to share attention with other elephants. Typically, however, other members of the herd were still present elsewhere in the same habitat at the time of testing, so they maintained access to conspecifics if they decided to terminate the session at any time. Because the Busch Gardens herd is managed in an open economy, an elephant could leave the testing area not only to interact with conspecifics, but also to access the waterfall, swimming pool, dirt pile, hay/browse, or any other part of the habitat. To maintain voluntary participation, it was up to the trainers who conducted the data collection to keep the session reinforcing while the elephant was tested.

Not only was participating in the trials one behavioral choice among many available in the environment, but the paradigm itself gave the elephant the ability to choose between two trays. As the alternate tray was removed as soon as the other was selected, the elephants learned quickly that they could not have both, which encouraged them to choose wisely. This challenge became a game or even a puzzle that provided mental stimulation. They had to gather information about the baited trays visually, consider their options, and then choose. The order of the conditions increased in difficulty as the elephants mastered each one, ensuring that the task continued to be stimulating throughout the course of data collection.

In addition, the sessions involved various physical activities. First, the elephants had to maneuver their trunk through the window in the husbandry wall to access the trays. Then, they had to use fine motor dexterity to manipulate the trays using the prehensile tip of their trunks. Finally, between each trial, they turned 180 degrees and walked between the testing area and the trainer waiting in a designated space. While these physical movements are already promoted with other, more traditional environmental enrichment designs and trainer interactions throughout the day, the research sessions proved to be an extension of that strategy.

The behaviors each elephant demonstrated while taking part in these research sessions would be considered species-specific. For example, elephants are known to be intelligent, large-brained mammals, and they were required to use those brains actively to succeed in the task. In addition, elephants use their trunk to reach and pick up desired items at a distance, which was inherent in the design of this study. Furthermore, while they were engaged fully with the trainers and task during trials, the elephants did not exhibit any stereotypical behaviors, as they are incompatible. Therefore, not only did participating in the research promote species-specific behaviors, but it also discouraged undesirable behaviors indicative of stress or boredom—another goal of enrichment.

There are challenges inherent in applied behavioral research, but they are not insurmountable. Most significantly, as any animal professional knows, the care and welfare of the animals are the top priority. Their well-being should never be compromised, including, but not limited to, extreme deprivation, forced participation, negative effects on social dynamics, or the use of aversive stimuli. Although following these guidelines may create some delays, prior creative and flexible planning can ensure a research project's success. For example, trainers in the means-end study had to be creative with the type of food used to bait the trays, using novelty, variety, and scarcity of the specific food type while still providing access to other food before trials. This ensured that their welfare was not compromised but participation was still likely.

Many factors are beyond the zoo staff/experimenters' control, including the weather, biological processes (e.g., behavior related to breeding), or even wildlife not managed by the facility (e.g., hungry squirrels stealing bait). The key is to identify when these issues can be worked around or simply when patience is required to wait out the situation. Because of the environmental variability in any applied study, it is desirable to control for as many potentially relevant factors as possible. Maintaining consistency by not running trials on rainy days, for example, helped rule out rain as a possible influence on each subject's behavior and performance. Furthermore, we took steps to control for unintentional trainer effects, such as keeping the trainer near the baited trays "blind" to which side was correct.

With strict protocols in place to control for extraneous factors, the trainers were still able to incorporate variability in the sessions in ways that would not affect the study. This played a crucial role in ensuring the task was sufficiently reinforcing to maintain the elephants' interest. Not only did the baited item vary from trial to trial, but the trainer who called away the elephant while the next trial was set up used a variety of rewards for coming to the trainer as well. To balance the reinforcement between the testing area and away space, this trainer took great care to use minimal primary reinforcers and incorporate numerous secondary reinforcers, including tactile stimulation and object retrievals, for example.

While the study described serves as a single example, the many other cognitive and behavioral research projects completed already, and those not yet imagined, provide infinite opportunities to enrich zoo and aquarium animals. We still have much to learn about animals' cognitive abilities, and conducting research creates more opportunities to change the environment, provide behavioral choices, promote species-specific behaviors, and enhance the welfare of zoo animals overall in the process. Although the argument posed here supplies only a subjective, anecdotal piece of evidence supporting its case, it can serve as a starting point for future research. Many animals of many species may appreciate the opportunity to participate in empirical tests of whether or not behavioral research actually is enriching.

FUTURE DIRECTIONS

We would like to encourage elephant teams to pursue research collaborations with local colleges and universities. Faculty and students can be a valuable resource in conducting cognitive research, from assisting with the logistics and data collection to providing fresh perspectives and new questions. It has been our experience that these zoo-university relationships provide bidirectional benefits, enriching both the students' education and the professionals' work environments (which, of course, also benefits the elephants). In his recent book, *Professor in the Zoo*, Maple (2016) testified about these relationships: "Once mutual benefits are explored by academic and zoo leaders, partnership is relatively easy to implement. To make it happen, both sides must acknowledge the value that each partner brings."

There was much discussion at the recent Elephant Wellness Conference about the way in which cognitive research could be integrated into existing programs. We encourage such discussion among elephant trainers to brainstorm ideas and develop interinstitutional comparisons. For example, we have now completed the means-end study at both Busch Gardens and Lowry Park Zoo, which allows for cross-species comparisons of Asian versus African elephants. It is important to note that replication in comparative psychology is very important. Most studies examining the cognitive abilities of animals are limited to a few individuals (Agrillo & Petrazzini, 2012). Therefore, replication is needed, especially when studying exotic species. In fact, Perdue et al.'s 2012 study on quantity judgment in elephants was a replication of an earlier study (Irie & Hasegawa, 2012). The replication revealed a different pattern than the one found in the original study. This type of discrepancy emphasizes the need for replications and interinstitutional comparisons.

Long term, we regard cognitive research as a bridge to improve communication between elephants and their human caretakers. We foresee the need for numerous studies that will continue to increase our understanding of elephant cognition. Better understanding of information-processing and other aspects of elephant cognition could help inform the design of a lexigram device that could allow us to "ask" elephants questions about their preferences systematically and scientifically. Thus, consistent with the spirit of the animal wellness movement, this form of enhanced communication would increase our knowledge while expanding zoo elephants' opportunities to make choices and exert control over their environment.

As humans and elephants experience increasing conflict in the wild, progressive zoos have an opportunity (some might say an obligation) to improve our understanding of elephant cognition. As world trends drive our species closer together, the preservation of elephants may depend upon comprehension of their needs derived from empirical research. Passive observation, long favored, will have to be augmented or replaced by active cooperation and eager engagement between humans and elephants. What better way is there to undertake this vital work than with guests watching, in a nurturing zoo environment that supports and shares the importance of elephant cognitive research?

FIGURE 1: MAKE NO "BUTTS" ABOUT IT: THE ELEPHANT INTERACTION AREA AT BUSCH GARDENS TAMPA BAY IS POPULAR WITH ELEPHANTS AND GUESTS ALIKE FOR ITS ABILITY TO DEMONSTRATE THE BENEFITS OF A PROGRAM COMMITTED TO POSITIVE REINFORCEMENT TRAINING (PHOTO: O. FAD).

FIGURE 2: BULL ELEPHANT PARTICIPATING IN COGNITIVE RESEARCH AT BUSCH GARDENS TAMPA BAY (PHOTO: J. SPENCER).

REFERENCES

Agrillo, C., & Miletto Petrazzini, M. E. (2012). The importance of replication in comparative psychology: the lesson of elephant quantity judgments. *Frontiers in Psychology, 3,* 181.

Anderson, U. S., Stoinski, T. S., Bloomsmith, M. A., Marr, M. J., Smith, A. D., & Maple, T. L. (2005). Relative numerousness judgment and summation in young and old Western lowland gorillas. *Journal of Comparative Psychology, 119*(3), 285.

Bowler, M. T., Buchanan-Smith, H. M., & Whiten, A. (2012). Assessing public engagement with science in a university primate research centre in a national zoo. *PLOS ONE, 7*(4), e34505.

Chevalier-Skolnikoff, S., & Liska, J. O. (1993). Tool use by wild and captive elephants. *Animal Behavior, 46,* 209–219.

Foerder, P., Galloway, M., Barthel, T., Moore, D. E. III, & Reiss, D. (2011). Insightful problem solving in an Asian elephant. *PLOS ONE, 6,* e23251.

Greco, B. J., Brown, T. K., Andrews, J. R. M., Swaisgood, R. R., & Caine, N. G. (2013). Social learning in captive African elephants (*Loxodonta africana africana*). *Animal Cognition, 16*(3), 459.

Highfill, L., Spencer, J. M., Fad, O., & Arnold, A. M. (2016). Performance on a means end task by Asian elephants (*Elephas maximus*) in a positive reinforcement-based protected contact setting. *International Journal of Comparative Psychology, 29*(1), 1-10. Retrieved from https://escholarship.org/uc/item/60z8438t

Irie, N., & Hasegawa, T. (2012). Summation by Asian elephants (*Elephas maximus*). *Behavioral Sciences, 2*(2), 50-56.

Irie-Sugimoto, N., Kobayashi, T., Sato, T., & Hasegawa, T. (2008). Evidence of means-end behavior in Asian elephants (*Elephas maximus*). *Animal Cognition, 11*, 359–365.

Maple, T. L. (2016). *Professor in the zoo: Designing the future for wildlife in human care.* Tequesta, FL: Red Leaf Press.

Meehan, C. L., & Mench, J. A. (2007). The challenge of challenge: Can problem solving opportunities enhance animal welfare? *Applied Animal Behavior Science, 102*(3), 246–261.

Neuringer, A. J. (1969). Animals respond for food in the presence of free food. *Science, 166*(3903), 399–401.

Perdue, B. M., Talbot, C. F., Stone, A. M., & Beran, M. J. (2012). Putting the elephant back in the herd: Elephant relative quantity judgments match those of other species. *Animal Cognition, 15*, 955–961.

Plotnik, J. M., de Waal, F. B. M., Moore, D., & Reiss, D. (2010). Self-recognition in the Asian elephant and future directions for cognitive research with elephants in zoological settings. *Zoo Biology, 29*, 179–191.

Plotnik, J. M., de Waal, F. B., & Reiss, D. (2006). Self-recognition in an Asian elephant. *Proceedings of the National Academy of Sciences, 103*, 17,053–17,057.

Plotnik, J. M., Lair, R., Suphachoksahakun, W., & de Waal, F. B. (2011). Elephants know when they need a helping trunk in a cooperative task. *Proceedings of the National Academy of Sciences, 108*(12), 5,116–5,121.

Poole, J. (1996). *Coming of age with elephants.* Chicago, IL: Trafalgar Square.

Rensch, B. (1956). Increase of learning capability with increase of brain-size. *American Naturalist, 90,* 81–95.

Ross, S. R. (2010). How cognitive studies help shape our obligation for ethical care of chimpanzees. In A. B. Smith (Ed.), *The mind of the chimpanzees: Ecological and experimental perspectives* (pp. 309–319). IL: University of Chicago Press.

Shoshani, J., Kupsky, W. J., & Marchant, G. H. (2006). Elephant brain part I: Gross morphology, functions, comparative anatomy, and evolution. *Brain Research Bulletin, 70,* 124–157.

Xitco, M. J., Gory, J. D., & Kuczaj, S. A., II. (1999). An introduction to the Living Sea's dolphin keyboard communication system. In K. Ramirez (Ed.), *Animal training: Successful animal management through positive reinforcement* (pp. 493–498). Chicago, IL: Shedd Aquarium Press.

Chapter 8

SENSORY ENVIRONMENTS OF ZOO ANIMALS: MEASURING ANTHROPOGENIC SOUND, LIGHT, AND GROUND VIBRATION IN A LARGE, MULTI-SPECIES EXHIBIT

Joseph Soltis, David Orban, Kimberly Adams, Jill Mellen, and Lori Perkins
Disney's Animals, Science and Environment, Disney's Animal
Kingdom®, Bay Lake, Florida 32830

Exhibit design involves more than the physical structure. Sound, illumination, vibration, air quality, and activity in the visual field are all aspects of an animal's sensory environment. Many stimuli, including the proximity of guests or animal care staff, features nearby such as food venues, entertainment, or construction, and more distant features such as vehicle traffic, air traffic, and industrial activity, may influence animals' sensory environments. At Disney's Animal Kingdom®, the very popular Kilimanjaro Safaris Ride allows guests to journey on "safari vehicles" that pass through or next to various exhibits displaying multiple African species, from small ungulates to elephants. Thus, the safari trucks alone may affect the sound and vibration levels that animals experience in their exhibits. In addition, the safari is illuminated for several hours at night to create the illusion of "sunset" and "twilight" that may produce light effects.

Therefore, we set out to measure and map visual representations of the sound, light, and seismic influences across these exhibits. Specifically, we created a "soundscape" that allows visualization of the sound pressure levels across exhibits attributable to the engine noise emanating from the safari vehicles. Second, we created a "lightscape" to illustrate the light intensity levels across one exhibit illuminated to create a "twilight" effect. Finally, we created a "vibration-scape" to depict the ground vibrations attributable to passing safari vehicles.

In this way, we sought to ensure that: 1) sound profiles across exhibits varied and included areas in which the ride vehicles had only slight influences; 2) light intensity profiles at night varied and included areas with low light effects, and 3) vibration profiles across exhibits varied and included areas of low vibration from the ride vehicles. While the process of ensuring proper sensory environments is ongoing, the results of these measurements have shown that exhibits have a variety of "micro-environments" with respect to sound, light, and vibration. These micro-environments reduce the potential effects of anthropogenic influences and provide the animals with the option to move to areas that are affected to a greater or lesser degree. These results showed the ways in which sensory environments can be quantified to allow zoo professionals to monitor and promote positive sensory environments for their animals.

EFFECTS OF ANTHROPOGENIC SOUND, LIGHT, AND VIBRATION

Sound

Populations of animals have evolved in and adapted to natural sound environments over millions of years. The effects on wild, laboratory, and zoo animals of human-generated (anthropogenic) sound have been reviewed widely (Nowacek, Thorne, Johnston, & Tyack, 2007; Morgan & Tromborg, 2007; Barber, Crooks, & Fristrup, 2010; Kight & Swaddle, 2011; Read, Jones, & Radford, 2014; Shannon et al., 2015). Anthropogenic sound has been demonstrated to have certain detrimental effects on marine, terrestrial, and aerial species. Noise can elicit a stress response in individuals (both behavioral and physiological), and can disrupt important behavioral processes, including vocal communication, foraging, mating, and rearing of offspring. Table 1 shows the range of known effects of anthropogenic noise on animals, and Table 2 shows some example effects of noise on animals specifically under human care.

Table 1

Some known effects of anthropogenic noise on animals

BEHAVIOR	Increased "abnormal" behavior (e.g., stereotypy)
	Increased vigilance and aggression
	Disruption of vocal communication
	Disruption of auditory detection of predators / prey
	Decreased breeding behavior and foraging behavior
	Desertion of offspring
	Disruption of learning processes
PHYSIOLOGY	Elevated stress hormones (e.g. cortisol, adrenaline)
	Hypertension
	Compromised immune response
	DNA damage, alteration of gene expression
	Temporary or permanent hearing threshold shifts
	Hearing loss
	Weight loss, loss of body condition
POPULATION	Decreased reproductive success
	Compromised reproduction (e.g. low birth weight, low egg production)
	Population decline, reduced abundance
	Species richness decline
Sources: Barber et al., 2010; Knight & Swaddle, 2011; Read et al., 2014; Shannon et al., 2015.	

Table 2

Examples of effects of various noise levels on animals in labs, zoos and aquariums

SPECIES	NOISE EXPOSURE: LEVELS**	RESPONSE TO INCREASED NOISE	
Shore crabs *Carcinus maenas*	Simulated ship noise: 148 - 155 dB ***	Decreased: feeding, predator avoidance	1*
Lined seahorses *Hippocampus erectus*	White noise: 130 dBZ tanks vs. 111 dBZ ***	Increased: distress behaviors, cortisol, parasitic infection Decreased: body weight	2
Anole lizards *Anolis spp.*	Human's loud voice: 74–76 peak dB (quiet) 95–105 peak dB (voice)	Increased: escape speed when threatened physically	3
Honeycreepers *Drepanididae*	Environmental noise: Normal vs. concert/ machine noise (no dB values)	Increased: corticoids Decreased: foraging and courtship	4
Gibbons *Hylobates lar*	Crowd noise: Low: 55-65 dB Medium: 65-70 dB High: >70 dB	Increased: scratching, open mouth display; looking at public, locomotion Decreased: grooming	5
Macaques *Macaca fascicularis*	Construction explosions: 43 dBA vs. 92 dBA	Increased: flight behavior, vocalizations, cortisol	6
Pandas *Ailuropoda melanoleuca*	Ambient noise: 65 dB vs. 72 dB	Increased: scratching; stress in voice; locomotion, cortisol	7
Rats *Rattus rattus*	White noise: 100 dB	Increased: neurological changes associated with Alzheimer's disease.	8

*1: Wale, Simpson, & Radford, 2013; 2: Anderson et al., 2011; 3: Jones & Jayne, 2012; 4: Shepherdson, Carlstead, & Wielebnowski, 2004; 5: Cooke & Schillaci, 2007; 6: Westlund et al., 2012; 7: Owen, Swaisgood, Czekala, Steinman, & Lindburg, 2004; 8: Sobrian et al., 1997.

** dBA = sound level weighted for human hearing; dBZ = no weighting; dB = weighting not specified.

*** Sound is measured on a logarithmic decibel (dB) scale, where 0 dB is referenced to a specific sound level. In air, the reference value is 20 μPa, the minimum sound audible to humans. In water, the reference value is 1 μPa, so these absolute dB values cannot be compared directly to those measured in air.

While absolute decibel (dB) levels are important, several caveats should be made. First, noise does not affect all species equally. For example, anthropogenic noise in urban environments can be at its highest levels during the day, and thus have a disproportionate influence on diurnal animals (Herrera-Montes & Aide, 2011), and species with different hearing ranges (e.g., infrasonic and ultrasonic hearing) may be affected uniquely by environmental sound levels of different frequencies (Morgan & Tromborg, 2007). Second, not all individuals react similarly to the same noise exposure. For example, Stoinski, Jaicks, & Drayton (2012) showed that individual gorillas reacted to increased crowd-size in different ways depending on group membership, sex, and personality rating. Third, sound quality is also important. Sounds that are unpredictable, harsh, and/or have a sudden onset can have negative effects, while predictable, normal, or soothing sounds can have neutral or positive effects on animals (Prior, 2006; Videan, Fritz, & Murphy, 2007; Morgan & Tromborg, 2007; Wells & Irwin, 2008; Westlund et al., 2012; Alworth & Buerkle, 2013).

Light

Light and light-dark cycles produced by our sun and more distant stars have shaped the adaptations of animal populations on Earth over millions of years. Humans have imposed new sources of artificial light that can affect animals profoundly. These effects can be particularly strong on animals under human care, where sources of artificial light compete with, or replace natural light.

Light-dark cycles: Animals have adapted to the light-dark cycles attributable to the Earth's revolutions and to seasonal shifts in day-length because of the Earth's orbit around the sun. These light signals control all of the internal biological rhythms of animals on a daily ("circadian") and annual basis (Bradshaw & Holzapfel, 2007; Goldman, 1999; Walton, Weil, & Nelson, 2011; Dahl & Thompson, 2012; Eisenstein, 2013; Peplow, 2013). Table 3 summarizes the many and pervasive biological processes that are entrained by light cycles, including activity budgets, sleep-wake cycles, body condition, and the function of reproductive, immune, digestive, and stress-response systems. Artificial light regimes may affect any of these processes. Table 4 provides further examples of the types of biological processes that can be affected by photoperiod manipulation.

Table 3

Common biological processes entrained to or affected by natural circadian light-dark cycles or natural seasonal changes in day-length in mammals and birds.

SLEEP-WAKE CYCLES AND ACTIVITY BUDGETS	Circadian sleep/wake cycles
	Circadian activity levels (e.g., foraging)
	Seasonal sleep/wake cycles (e.g., hibernation)
	Seasonal activity levels
	Seasonal migration
	Lifespan
BEHAVIORAL EXPRESSION OF EMOTION AND COGNITION	Seasonal changes in "anxious," "depressive," and aggressive behaviors
	Seasonal changes in spatial learning ability
BODY CONDITION	Seasonal changes in growth, fat storage, pelage condition
	Circadian and seasonal changes in metabolism
REPRODUCTIVE FUNCTION	Skeletal and eye morphology
IMMUNE FUNCTION	Immunological defense (e.g., immune cell number, antibody response)
DIGESTIVE FUNCTION	Seasonal changes in intestinal absorption
	Seasonal changes in intestinal bacterial communities.
STRESS RESPONSE	Circadian glucocorticoid secretion (e.g., just prior to active phase of the day)Seasonal glucocorticoid secretion (e.g., in breeding season)
	Adrenal response

Sources: Goldman, 1999; Bradshaw & Holzapfel, 2007; Morgan & Tromborg, 2007; Walton et al., 2011; Dahl & Thompson, 2012; Dickmeis et al., 2013; Peplow, 2013; Manser, 1996.

Table 4

Some biological processes affected by shifting light-dark cycles (L:D) in mammals

SPECIES	LIGHT CONDITIONS	EFFECTS	
Rats *Rattus norvegicus*	Change from 12:12 to 8:16 L:D cycle	Short day length: Cellular instability (pre-cancer)	1
Sheep (Soay breed)	Long day length (16:8) vs. short day length (8:16)	Short day length: Decreased: appetite-related peptides, food intake, rumen bacterial diversity Increased: sex	2
Mice *Mus musculus*	Continuous light vs. 12:12 L:D cycle	Continuous light: Increased: corticosterone, latency to aggression	3
Boars (several breeds)	Natural changes in L:D cycles	Shortening day lengths: Increased: sperm volume and motility	4
Mouse lemurs (*Microcebus murinus*)	Long (14:10) vs. short day (10:14)	Short day length: Increased: food intake, fat storage Cessation: sexual activity	5
In utero cavies (*Cavia aperea*)	Increasing vs. decreasing day length during gestation	Shortening day lengths: Increased: age at maturity Decreased: exploration	6
Siberian hamsters (*Phodopus sungorus*)	Long day length (16:8) vs. short day length (8:16)	Short day length: Increased: allergic response Decreased: size, reproductive tissue	7
Mouse lemurs (*Microcebus murinus*)	Artificial seasons (> 1 cycle/yr)	Accelerated "annual" cycles: Decreased: life-span	8
Collared lemmings (*Dicrostonyx groenlandicus*)	Long (22:2), intermediate (16:8), vs. short (8:16) day lengths.	Shorter day lengths: Increased: winter pelage coloration; male gonadal regression, anxiety- and depressive-like behaviors	9

Brandt's voles (*Microtus brandti*)	Long (16:8) vs. short (8:16) day length	Short day length: Increased: basal metabolic rate, non-shivering (physiological) thermogenesis, Decreased: body weight	10
Sources: 1: Adam et al., 2011; 2: Clarke, Rao, Chilliard, Delavaud, & Lincoln, 2003; McEwan et al., 2005; 3: Van der Meer, Van Loo, & Baumans, 2004; 4: Knecht, Środoń, Szulc, & Duziński, 2013; 5: Perret & Schilling, 1995; Génin & Perret, 2000; 6: Guenther & Trillmich, 2012; 7: Martin & Coon, 2012; 8: Perret, 1997; 9: Weil, Bowers, & Nelson, 2007; 10: Zhao & Wang, 2005.			

Quality of light (wavelength): Humans can perceive only a small portion of the electromagnetic spectrum. For example, many rodents, birds, reptiles, and invertebrates can perceive ultraviolet light, which is invisible to humans (Davies, Benni, Inger, Hempel de Ibarra, & Gaston, 2013). In an artificial light environment, it is necessary to provide all wavelengths (including UV light) that are necessary for a particular species' proper biological functioning (e.g., visual foraging, mate choice, etc.), or that animals prefer (Honkavaara, Koivula, Korpimäki, Siitari, & Viitala, 2002; Prescott, Wathes, & Jarvis, 2003; Morgan & Tromborg, 2007; Ross, Gillespie, Hopper, Bloomsmith, & Maple, 2013). Where light is considered "pollution" (e.g., nighttime light that interferes with sleep), then narrowing the wavelengths transmitted selectively can reduce the effects. In humans, for example, artificial light can interfere with natural biological rhythms, such as normal sleep cycles, but visible light with longer wavelengths (e.g., reds) has a much less disruptive effect compared to that of mid and short wavelengths (e.g., greens and blues: Morita & Tokura, 1998; Brainard et al., 2008).

Light intensity: Animals differ in their ability to detect and perceive light intensity. For example, dogs, cats, and birds are more sensitive to light than are humans (Prescott & Wathes, 1999; Morgan & Tromborg, 2007), and too much light can have negative effects on them. Further, nighttime light "pollution" can disrupt circadian processes in humans (e.g., Morita & Tokura, 1998; McColl & Veitch, 2001; Stevens, 2006; Dijk et al., 2013; Benke & Benke, 2013). This effect has been linked to a range of behavioral issues, such as disruption of sleep cycles, and reduced alertness and performance. It also has been linked to illnesses, including 1) breast cancer; 2) prostate cancer; 3) heart disease; 4) obesity, and 5) diabetes (Dominoni, Quetting, & Partecke, 2013. 2: Kempenaers, Borgström, Loës, Schlicht, & Valcu, 2010. 3: Frank, Evans, & Gorman, 2010, 4: Le Tallec, Perret, & Théry, 2013. 5: Stone, Jones, & Harris, 2009). Therefore, nighttime light "pollution" ideally should be of low intensity to reduce its effects.

Vibration

Just as animal populations have evolved and adapted to their sound and light environments, so have they adapted to the physical surroundings in which they live. Many animals glean information from substrate vibrations, which can mediate predator-prey relationships as well as social communication (e.g., Narins, Lewis, Jarvis, & O'Riain, 1997; Hill, 2001; Caldwell, McDanil, & Wakentin, 2010). Anthropogenic vibration, on the other hand, is prevalent in human-dominated landscapes, and may affect both people and other animals (e.g., Norton, Kinard, & Reynolds, 2011). For example, vibration from trains can disrupt sleep and increase heart rate in humans (Smith, Croy, Ögren, & Waye, 2013, 2016), and walking within 6 m of subterranean mole rats elicited retreat behavior (Šklíba, Šumbera, & Chitaukali, 2008). In piglets, whole body vibrations that mimicked transport resulted in less time spent lying down, and increased levels of stress hormones (Perremans, Randall, Rombouts, Decuypere, & Geers, 2001). In zoos, construction has been shown to affect animal behavior and physiology (e.g., Powell, Carlstead, Tarou, Brown, & Montfort, 2006; Chosy, Wilson, & Santymire, 2014; Boyle et al., 2015). Although airborne noise was the focus of these investigations, it is possible that in-ground vibration also affects animal behavior or physiology during construction events.

METHODS: PHYSICAL MEASUREMENT OF SOUND, LIGHT, AND GROUND VIBRATION

Soundscape

We created a "soundscape" that depicted the expected sound pressure levels from safari vehicles across the exhibits on the Kilimanjaro Safaris Ride. Sound level measurements were collected along a road perpendicular to the ride path at various distances from it (13-125 ft). The measurements were taken between exhibits in a non-animal area. Sound pressure level (SPL; re: 20 μPa) was measured with a 3M™ Sound Examiner SE402 (3M, St. Paul, MN, USA) sound pressure level meter (C-weighting, slow response) at a height of 45 in from the ground. We chose a C-weighting to capture low frequencies that many of our larger animals are likely to perceive, rather than an A-weighting, which is weighted for human perception. The meter was set to log the maximum (*Lcsmx*) and mean (*Leq*) dBC levels every 5 seconds. By combining all sound level measurements of passing trucks at different distances, a pattern emerged that showed the way in which SPL decreased with distance from the source (Fig 1). Best-fit logarithmic equations were computed in Excel for the recorded mean and max values. This allowed extrapolation of sound pressure levels at other, unmeasured distances across the landscape.

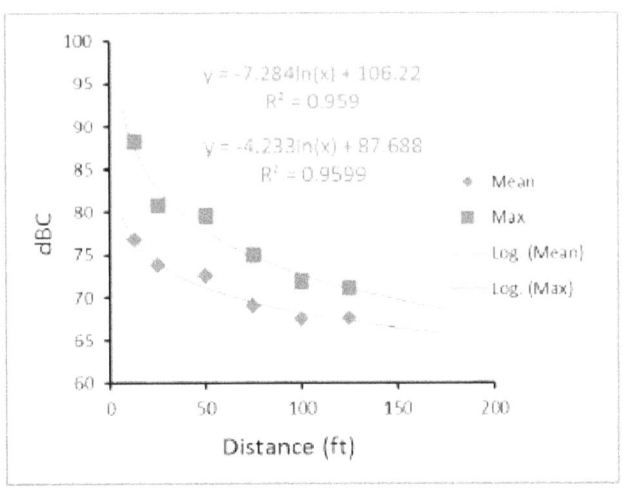

FIG. 1. *EMPIRICAL DATA ON SOUND PRESSURE LEVELS FROM SAFARI VEHICLES AT DIFFERENT DISTANCES FROM THE RIDE PATH. FOR TRUCKS PASSING BY, THE MAXIMUM AND MEAN DBC VALUES ARE SHOWN FOR DIFFERENT DISTANCES FROM THE RIDE PATH. LOGARITHMIC TREND LINES ARE SHOWN, TOGETHER WITH EQUATIONS TO EXTRAPOLATE TO UNMEASURED DISTANCES, AND R^2 VALUES SHOW THE VARIANCE EXPLAINED. THE MAX EQUATION (IN RED) WAS USED TO CREATE THE SOUNDSCAPE SHOWN IN FIG. 4.*

Soundscape maps were developed in ArcMap (ESRI® ArcGIS v10.0) using the Multi-Ring Buffer Tool to create buffer rings at 10 ft increments around the KSR vehicle path and give a high resolution depiction of the sound gradient. Estimated max dBC sound levels were calculated for each buffer ring using the best-fit max logarithmic equation (Fig. 1) with the mid-distance of each ring used as the input x-value (i.e., 5 ft for the 0–10 ft ring). The rings were then trimmed to the boundaries of the exhibits using the Clip Tool to focus on what the animals experience in their exhibits. Max dBC values across the soundscape represent the maximum sound level experienced by an animal as a KSR truck passes by.

Lightscape

We created a "lightscape" that depicted the light intensity levels from nighttime artificial lighting on the Kilimanjaro Safaris Ride. The lighting regime across the entire landscape is ongoing except for one exhibit that is complete (the "Finale" exhibit), so we created a lightscape for this exhibit. A grid of points at 50 ft intervals was overlaid (ESRI® ArcGIS v10.0) on the exhibit map, and we measured light intensity in the exhibit at each of those points (Fig. 2). Note that 3 points were moved when measuring the exhibit: 2 points in water features, and one on the road with poor visibility of oncoming vehicular traffic.

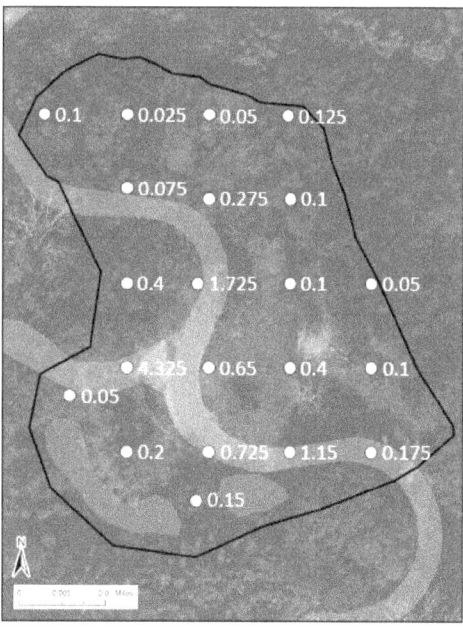

Fig. 2. *Empirical data on artificial light intensity across an animal exhibit. Ride path shown in tan and water features shown in blue. The mean light intensity (in lux) is shown at each point. These data were used to create the lightscape shown in Fig. 5.*

Light illuminance (in lux, *lx*) was measured with a LI-COR® LI-250A light meter (LI-COR, Lincoln, NB, USA), using a LI-210R photometric sensor. To measure the animal's light intensity experience at each location, the light meter was placed on a monopod 66 in high. This height was chosen as a rough center point given the potentially short and tall African species that could be in the exhibit. At each location, a 15 s average was taken horizontally in each of the cardinal directions (N, S, E, and W). The average experience for that point was calculated as the mean of the 4 readings. This averaging method obscured the highest and lowest possible light intensities that the animals could experience, but the effect was small. The point with the highest intensity showed the greatest spread around the mean value of 4.33 lux, and the highest value at that point was 11.8 lux. However, the difference between these values was actually rather small. The average value reflects light between twilight and a clear night with a full moon, and the highest value reflects twilight light. Therefore, the mean at each location provided a good representation of the animals' light experience at that location.

The lightscape map was developed in ArcMap (ESRI® ArcGIS v10.0) using the Spline Interpolation Tool to extrapolate from the points measured in the grid to unmeasured areas across the exhibit. The interpolated light gradient was then trimmed to the boundaries of the exhibits using the Clip Tool. The resulting lightscape represents the visual experience of an animal attributable to the artificial lighting in the exhibit.

Vibration-scape

We created a "vibration-scape" that illustrated the amplitudes of ground vibrations from safari vehicles across the exhibits on the Kilimanjaro Safaris Ride. Ground vibration measurements were collected in 3 areas (in the soil along two roads intersecting the ride path, and in the soil of the "Finale" exhibit) at various distances from the ride path (5-125 ft). Peak particle velocity (PPV) was measured with a MINI-SEIS III sensor (White Industrial Seismology, Inc., Joplin, MO, USA) using the geophone coupling spikes driven into the ground at each measuring location. The sensor was set to sample along each of three axes at 1 s intervals in Histogram mode and measure PPV in mm/s. One axis was perpendicular to the ride path, one parallel, and one vertical. The maximum amplitude of the three axes (PPV) during each 1 s sampling period was used for analysis. By combining all vibration measurements for passing trucks, a pattern emerged that showed the way in which vibration amplitude (PPV) decreased with distance from the source (Fig. 3). Best-fit logarithmic equations were computed in Excel for the recorded mean and max values. This allowed extrapolation of PPV at other, unmeasured distances across the KSR landscape.

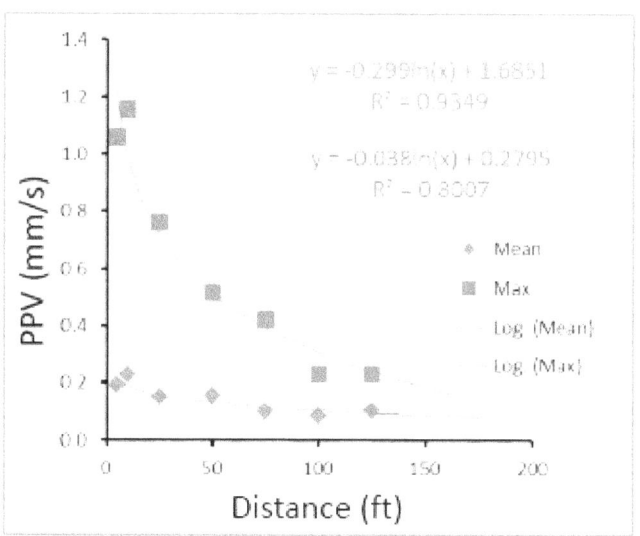

FIG 3. *EMPIRICAL DATA ON GROUND VIBRATION AMPLITUDE FROM SAFARI VEHICLES AT DIFFERENT DISTANCES FROM THE RIDE PATH. THE MAXIMUM AND MEAN VALUES OF THE PEAK PARTICLE VELOCITY (PPV, MM/S) FOR PASSING TRUCKS ARE SHOWN FOR DIFFERENT DISTANCES. LOGARITHMIC TREND LINES ARE SHOWN, TOGETHER WITH EQUATIONS TO EXTRAPOLATE TO UNMEASURED DISTANCES, AND R^2 VALUES TO SHOW VARIANCE EXPLAINED. THE MAX EQUATION (IN RED) WAS USED TO CREATE THE VIBRATION-ESCAPE SHOWN IN FIG. 6.*

Vibration-scape maps were developed in ArcMap (ESRI® ArcGIS v10.0) using the Multi-Ring Buffer Tool to create buffer rings at 10 ft increments around the KSR vehicle path and give a high resolution depiction of the vibration gradient. Estimated vibration amplitudes were calculated for each buffer ring using the best-fit max logarithmic equation with the mid distance of each ring used as the input x-value (i.e., 5 ft for the 0-10 ft ring). The rings were then trimmed to the boundaries of the exhibits using the Clip Tool to focus on what the animals experience in their exhibits. The vibration amplitudes across the exhibit represented the maximum ground vibration level experienced by an animal as a KSR truck passes by.

RESULTS

Anthropogenic sound from the Kilimanjaro Safaris Ride at Disney's Animal Kingdom®

We measured the sound pressure levels from passing safari trucks at varying distances from the safari ride path. These measurements allowed us to estimate the maximum sound levels that animals experience from trucks passing alongside or across their exhibits. Figure 4 shows the resulting "soundscape" for all exhibits along the Kilimanjaro Safari Ride. These estimates suggested that when animals are in very close proximity to the ride path (5 ft), maximum sound pressure levels from passing trucks is 94 dBC. However, these levels drop off very quickly with distance from the source because of spreading loss and other attenuation factors. In fact, at just 15 ft, the estimated sound level was 86 dBC. As such, even in smaller exhibits, there are areas where exposure levels are substantially lower, with the trucks' sound often blending into ambient background sound levels in the 60s dBC range.

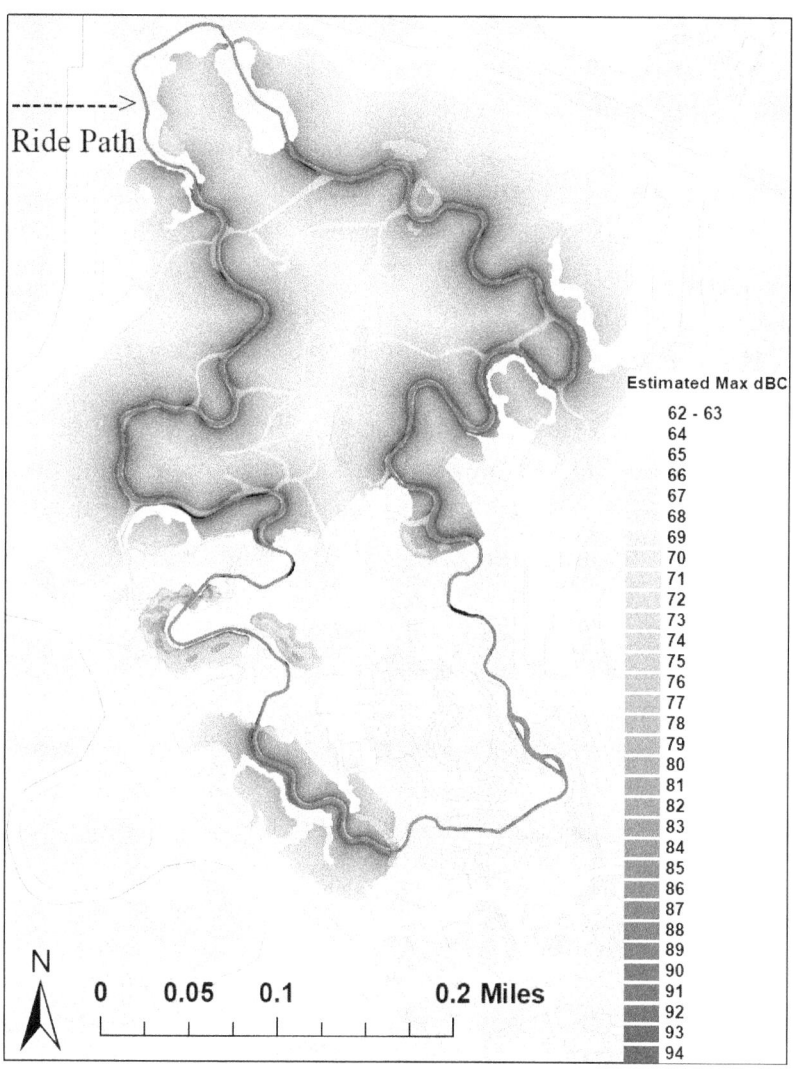

Fig 4. Soundscape depicting estimated maximum sound pressure levels (*d*BC) from safari trucks across animal exhibits. The arrow indicates the vehicle ride path (black line). Estimated sound pressure levels are shown in animal exhibits, and ranged from the low 60s *d*BC (in blue), when distant from the ride path, to the mid 90s *d*BC (in red) when very close to the ride path.

To put these sound pressure level values into perspective, they can be compared to Occupational Safety and Health Administration examples of sound pressure levels: quiet urban residence ≈ 50s dBA; human conversation ≈ 60s dBA; vacuum cleaner at 3 m ≈ 70s dBA; truck at 15 m ≈ 80s dBA; jackhammer at 15 m ≈ 90s dBA; discotheque club ≈ 110s dBA; jet takeoff at 100 m ≈ 120s dBA (OSHA, 1981).

The estimated maximum sound level animals experience because of the safari trucks may be relatively loud when in very close proximity to the ride path (5 ft or less). However, in exhibits where it is possible for animals to be that close to ride vehicles, the drivers are trained to slow down or stop for animals nearby, so the actual engine sound levels are likely lower than the estimates indicated, which were based on vehicle speeds between exhibits. Note also that these maximum sound pressure levels represent the brief, highest peak in amplitude as the truck passes, not the average of the sound as the truck approaches and then leaves, which would be lower. The most important result, however, is that the sound level diminishes quickly with distance from the ride path and can be as low as a human conversation when animals are quite distant, thereby allowing them to choose to move to quieter locations.

The sound pressure level gradient depicted in the soundscape in Fig. 4 is an estimate based on measurements in a sample location, and actual sound levels from the trucks may be different, depending on their operation (e.g., speed, acceleration) and the environment (e.g., topography, vegetation). Therefore, if there was an area of special interest, specific measurements would need to be conducted in that area. Nevertheless, such estimated soundscapes can be useful tools to explore the sound environments of zoo animals visually, and can support efforts to create and maintain positive sound environments. In this case, we determined that the animal exhibits showed great variation in the magnitude of the effect of trucks, such that animals had the choice to move to noisier or quieter areas according to their preferences.

When interpreting exhibit sound levels, it is best to consider the hearing ranges of specific animals, which often differ from humans. For example, tests of an Asian elephant (*Elephas maximus*) indicated a hearing range of 17 – 10,500 Hz at 60 dB (reviewed in Soltis, 2010), while the human hearing range is 31 – 17,600 Hz at 60 dB (see Heffner, 2004). On the high end, the galago (*Galago senegalensis*) hearing range is 92 – 65,000 Hz at 60 dB (Heffner, 2004). Therefore, to better understand the effect of noise levels on specific animals, the hearing range of affected species needs to be investigated and compared to the frequency profile of the sound source. The hearing range of exotic animals is not always known, however, and specialized equipment is required to measure amplitude at different frequencies. Nevertheless, the quality of sound and the unique sensitivities of species and individuals needs to be taken into account when assessing effect of noise on animal welfare.

Anthropogenic light in Kilimanjaro Safaris Ride exhibits at Disney's Animal Kingdom®

We measured the light intensity from artificial lighting at different points throughout the "Finale" exhibit of the Kilimanjaro Safaris Ride. These measurements allowed us to estimate the light intensities that animals experience from the artificial lights installed within the exhibit. Figure 5 shows the resulting "lightscape" for the exhibit. These results indicated that when animals are in the brightest area of the exhibit, they experience maximum light intensities of approximately 4 lux (maximum = 4.33 lux). However, light intensity varies considerably across the exhibit, with some areas showing substantially lower light intensities near zero lux (minimum = 0.03 lux).

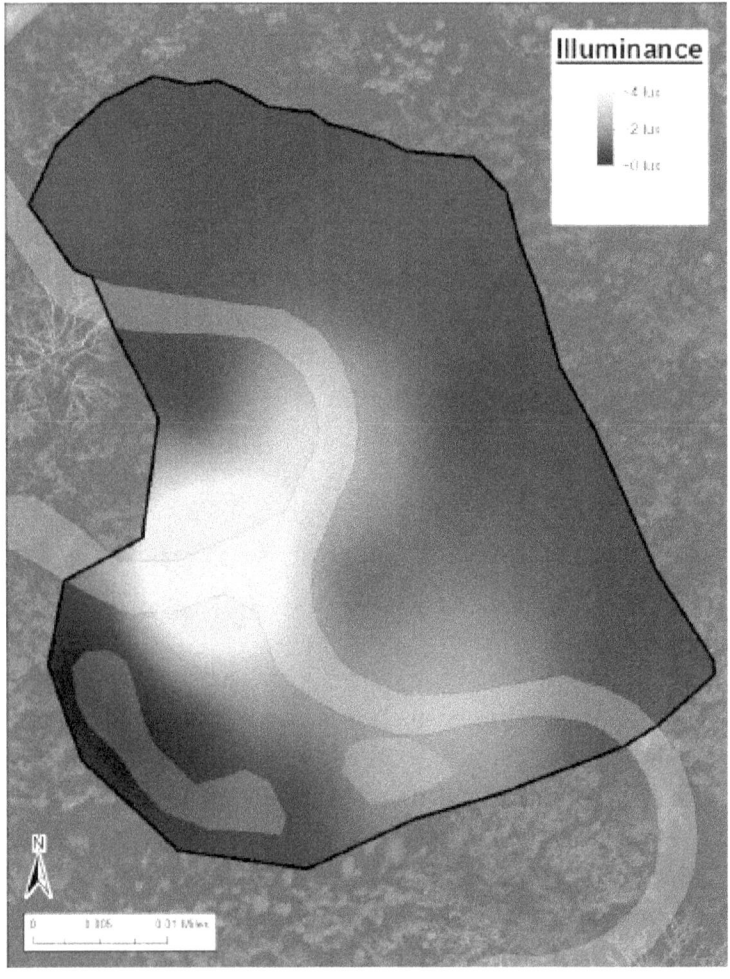

Fig 5. *Lightscape illustrating light intensity levels from artificial night lighting in the "Finale" exhibit. Ride path shown in tan and water features in blue. Light intensity levels ranged from 0.03–4.33 lux.*

To put these light intensities into perspective, they can be compared to examples of known light intensities: overcast night ≈ 0.0001 lux; full moon ≈ 0.5 lux; twilight ≈ 10 lux; office ≈ 250 lux; overcast daytime sky ≈ 1000 lux; direct sun clear sky ≈ 100,000 lux (Prescott et al., 2003).

Thus, the brightest parts of the exhibit are between a full moon and twilight, and the darkest areas are between an overcast night sky and a clear night with a full moon.

The production of such "lightscapes" can be a useful tool to examine zoo animals' light environments visually, and can support efforts to create and maintain positive light environments. In this case, we determined that the light intensity levels in animal exhibits varied significantly, and that the absolute values were not very high, thus providing an environment with medium to no effects from artificial light in which animals could choose to be in areas they preferred.

When interpreting exhibit light intensity levels, it is best to consider of the specific visual capabilities of animals, which often differ from humans. For example, the vision of elephants (*Loxodonta africana*) is adapted to both day and night (Yokoyama, Takenaka, Agnew, & Shoshani, 2005). Thus, elephants are capable of dichromatic vision (humans are trichromatic) during the day, and have greater sensitivity at night, compared to humans. Therefore, to better understand the effect of anthropogenic light on specific animals, the visual capacities of affected species need to be investigated and compared to the wavelength and intensity characteristics of the light source. The visual capacities of exotic animals are not always known, however, and specialized equipment is required to measure light intensity at different wavelengths. Nevertheless, the quality of light and the unique sensitivities of species and individuals needs to be taken into account when assessing effect of artificial light on animal welfare.

Anthropogenic ground vibration from the Kilimanjaro Safaris Ride at Disney's Animal Kingdom®

We measured the amplitudes of ground vibration from passing safari trucks at varying distances from the ride path. These measurements allowed us to estimate the maximum ground vibration that animals experience from trucks passing through or near their exhibits. Figure 6 shows the resulting "vibration-scape" for all exhibits along the Kilimanjaro Safaris Ride. These results showed that when animals are in very close proximity to the ride path (5 ft), maximum vibration (peak particle velocity) was approximately 1 mm/s. However, the vibration amplitude diminished very quickly with distance from the source, because of attenuation of vibration in the substrate. As such, even in smaller exhibits, most areas showed substantially lower vibration from the trucks that dropped almost to zero.

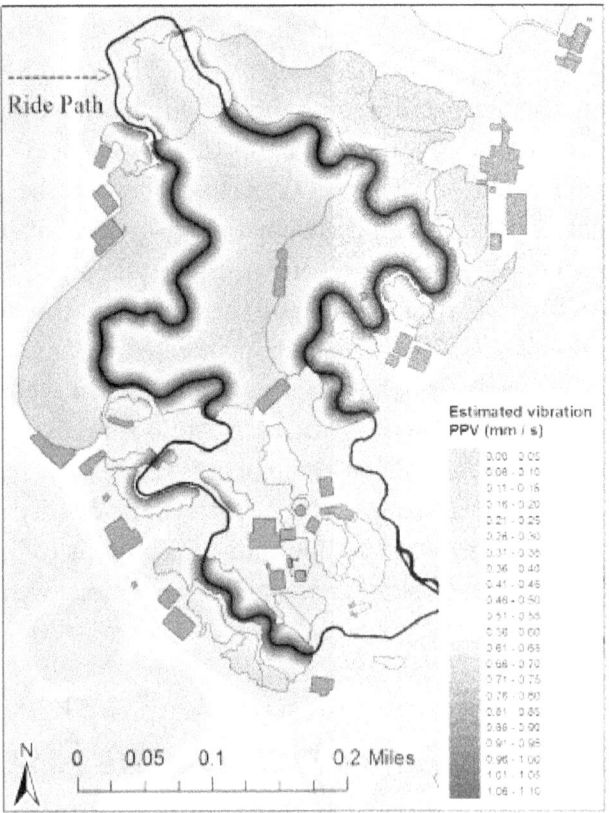

FIG 6. *VIBRATION-SCAPE SHOWING ESTIMATED SEISMIC VIBRATIONS (PEAK PARTICLE VELOCITY, PPV, MM/S) FROM SAFARI TRUCKS IN OR NEAR ANIMAL EXHIBITS. THE ARROW INDICATES THE VEHICLE RIDE PATH (BLACK LINE). ESTIMATED PPVS ARE SHOWN IN ANIMAL EXHIBIT AREAS, AND RANGED FROM ALMOST 0 MM/S (NO VIBRATION) WHEN DISTANT FROM THE RIDE PATH, TO APPROXIMATELY 1 MM/S WHEN VERY CLOSE TO IT.*

To put these vibration values into perspective, we can compare them to California Department of Transportation data on human perception of transportation and construction vibration: barely perceptible ≈ 1 mm/s; distinctly perceptible ≈ 6 mm/s; strongly perceptible ≈ 23 mm/s (Caltrans, 2013). We also performed an experiment by dropping a 15-pound weight (bowling ball) onto the ground (turf and soil) from a height of 3 ft, and measured peak particle velocity at different distances. The results were: 90 ft ≈ 0.25 mm/s; 60 ft ≈ 0.5 mm/s; 40 ft ≈ 1.0 mm/s; 30 ft ≈ 1.5 mm/s; 20 ft ≈ 2.0 mm/s; 10 ft ≈ 2.5 mm/s. Thus, the lowest vibration values in the exhibits are equivalent to "background" vibration (i.e., no vibration), and the majority of the values would be imperceptible to humans. The highest values, only in very close proximity to the ride path, are approximately the equivalent of dropping a bowling ball onto a lawn at 40 ft, and likely would be barely perceptible to humans.

Note that the vibration level gradient shown in the vibration-scape in Fig. 6 is an estimate based on measurements in sample locations, and actual vibrations from the trucks may differ, depending on their conditions (e.g., shocks, acceleration) and the environment (e.g., topography; substrate; vegetation, ride path condition). Therefore, if there was an area of special interest, specific measurements would need to be conducted in that area. In this case, we determined that the animal exhibits showed great variation in vibration from the passing trucks, with vibrations typically negligible unless within 10 ft of the truck, and even then, they were not substantial.

When interpreting vibration levels, it is best to consider the sensing capacities of the specific animals, which often differ from humans. For example, there is speculation that African elephants gather information from ground vibration (e.g., O'Connell-Rodwell et al., 2006), and the nocturnal, functionally blind Namib Desert golden mole (*Eremitalpa granti*) can detect termite prey solely from seismic cues (Narins et al., 1997). Therefore, to better understand the effect of vibration on specific animals, the characteristics of the vibration source, and the sensing modalities of affected animals need to be compared, with special attention to species which rely on vibration to gather information from their environment.

CONCLUSION

Zoo Animals' Sensory Environments

Zoo professionals face increasing challenges to design exhibits and animal care programs that both improve animal welfare and enhance the visitor experience. Therefore, they are looking for ways to make evidence-based improvements to animal welfare when designing or modifying exhibits. Ensuring proper sensory environments is an important part of this mission. In this paper, we described the ways in which the sensory environments of animals in zoo exhibits can be measured and represented visually. We created zoo "soundscapes," "lightscapes," and "vibration-scapes" that represented the animals' daily experiences of sound, light, and ground vibration. Our goal in each case was to ensure that these sensory effects varied within animal exhibits, such that animals are able to choose to move between areas of high and low or no influence. Such considerations should help promote the ethical management of animals wherever they are kept under human care.

REFERENCES

Adam, M. L., Torres, M. F. P., Franci, A. C., Sponchiado, G., Torres, R. A., & Correia, M. T. S. (2011). On the stress by photoperiod, temperature and noise as possible causes of genomic damaging in an animal model. *Stress and Health, 27*, e152-156.

Alworth, L. C., & Buerkle, S. C. (2013). The effects of music on animal physiology, behavior and welfare. *Lab Animal, 42*, 54-61.

Anderson, P. A., Berzins, I. K., Fogarty, F., Hamlin, H. J., & Guillette Jr., L. J. (2011). Sound, stress, and seahorses: The consequences of a noisy environment to animal health. *Aquaculture, 311*, 129-138.

Barber, J. R., Crooks, K. R., & Fristrup, K. M. (2010). The costs of chronic noise exposure for terrestrial organisms. *Trends in Ecology and Evolution, 25*, 180-189.

Benke, K. K., & Benke, K. E. (2013). Uncertainty in health risks from artificial lighting due to disruption of circadian rhythm and melatonin secretion: A review. *Human and Ecological Risk Assessment, 19*, 916-929.

Boyle, S. A., Roberts, B., Pope, B. M., Blake, M. R., Leavelle, S. E., Marshall, J. J., . . . Kouba, A. J. (2015). Assessment of flooring renovations on African elephant (*Loxodonta africana*) behavior and glucocorticoid response. *PLoS ONE, 10*, e0141009. doi: 10.1371/journal.pone.0141009

Bradshaw, W. E., & Holzapfel, C. M. (2007). Evolution of animal photoperiodism. *Annual Review of Ecology, Evolution, and Systematics, 38*, 1-25.

Brainard, G. C., Sliney, D., Hanifin, J. P., Glickman, G., Byrne, B., Greeson, J. M., . . . Rollag, M. D. (2008). Sensitivity of the human circadian system to short-wavelength (420-nm) light. *Journal of Biological Rhythms, 23*, 379-386.

Caldwell, M. S., McDanil, J. G., & Wakentin, K. M. (2010). Is it safe? Red-eyed treefrog embryos assessing predation risk use two features of rain vibrations to avoid false alarms. *Animal Behaviour, 79*, 255-260.

Caltrans. (2013). Transportation and construction vibration guidance manual. http://www.dot.ca.gov/hq/env/noise/pub/TCVGM_Sep13_FINAL.pdf (accessed August 19, 2016)

Chosy, J. C., Wilson, M., & Santymire, R. (2014). Behavioral and physiological responses in felids to exhibit construction. *Zoo Biology, 33*, 267-274.

Clarke, I. J., Rao, A., Chilliard, Y., Delavaud, C., & Lincoln, G. A. (2003). Photoperiod effects on gene expression for hypothalamic appetite-regulating peptides and food intake in the ram. *American Journal of Physiology–Regulatory, Integrative and Comparative Physiology, 284*, R101-R115.

Cooke, C. M., & Schillaci, M. A. (2007). Behavioral responses to the zoo environment by white handed gibbons. *Applied Animal Behaviour Science, 106*, 125-133.

Dahl, G. E., & Thompson, I. M. (2012). Effects of photoperiod on domestic animals. In R. J. Collier & J. L. Collier (Eds.), *Environmental physiology of livestock* (pp. 229-242). New York, NY: John Wiley & Sons, Inc.

Davies, T. W., Bennie, J., Inger, R., Hempel de Ibarra, N., & Gaston, K. J. (2013). Artificial light pollution: Are shifting spectral signatures changing the balance of species interactions? *Global Change Biology, 19*, 1417-1423.

Dijk, D. J., Duffy, J. F., Silva, E. J., Shanahan, T. L., Boivin, D. B., & Czeisler, C. A. (2013). Amplitude reduction and phase shifts of melatonin, cortisol and other circadian rhythms after a gradual advance of sleep and light exposure in humans. *PLoS ONE, 7*, e30037. doi:10.1371/journal.pone.0030037

Dominoni, D. M., Quetting, M., & Partecke, J. (2013). Long-term effects of chronic light pollution on seasonal functions of European blackbirds (*Turdus merula*). *PLoS ONE, 8*, e85069. doi:10.1371/journal.pone.0085069

Eisenstein, M. (2013). Stepping out of time: How can people better adapt to an "unnatural" world of artificial lighting and alarm clocks? *Nature, 497*, S10-S12.

Frank, D. W., Evans, J. A., & Gorman, M. R. (2010). Time-dependent effects of dim light at night on re-entrainment and masking of hamster activity rhythms. *Journal of Biological Rhythms, 25*, 103-112.

Génin, F., & Perret, M. (2000). Photoperiod-induced changes in energy balance in gray mouse lemurs. *Physiology and Behavior, 71*, 315-321.

Goldman, B. D. (1999). The circadian timing system and reproduction in mammals. *Steroids, 64*, 679-685.

Guenther, A., & Trillmich, F. (2012). Photoperiod influences the behavioral and physiological phenotype during ontogeny. *Behavioral Ecology, 24*, 402-411.

Heffner R.S. (2004). Primate hearing from a mammalian perspective. *The Anatomical Record Part A*, 281A, 1111-1122.

Herrera-Montes, M. I., & Aide, T. M. (2011). Impacts of traffic noise on anuran and bird communities. *Urban Ecosystems, 14*, 415-427.

Hill, P. S. M. (2001) Vibration and animal communication: A review. *American Zoologist, 41*, 1135-1142.

Honkavaara, J., Koivula, M., Korpimäki, E., Siitari, H., & Viitala, J. (2002). Ultraviolet vision and foraging in terrestrial vertebrates. *Oikos, 98*, 505-511.

Jones, Z. M., & Jayne, B. C. (2012). The effects of sound on the escape locomotor performance of anole lizards. *Journal of Herpetology, 46*, 51-55.

Kempenaers, B., Borgström, P., Loës, P., Schlicht, E., & Valcu, M. (2010). Artificial night lighting affects dawn song, extra-pair siring success, and lay date in songbirds. *Current Biology, 20*, 1735-1739.

Kight, C. R., & Swaddle, J. P. (2011). How and why environmental noise impacts animals: An integrative, mechanistic view. *Ecology Letters, 14*, 1052-1061.

Knecht, D., Środoń, S., Szulc, K., & Duziński, K. (2013). The effect of photoperiod on selected parameters of boar semen. *Livestock Science, 157*, 364-371.

Le Tallec, T., Perret, M., & Théry, M. (2013). Light pollution modifies the expression of daily rhythms and behavior patterns in a nocturnal primate. *PLoS ONE, 8*, e79250. doi:10.1371/journal.pone.0079250

Martin, L. B., & Coon, C. A. C. (2012). Photoperiod-driven variation in an allergic response is independent of allergen exposure. *Canadian Journal of Zoology, 90*, 1086-1093.

McColl, S. L., & Veitch, J. A. (2001). Full-spectrum florescent lighting: A review of its effects on physiology and health. *National Research Council Canada* (NRCC-43097).

McEwan, N. R., Abecia, L., Regensbogenova, M., Adam, C. L., Findlay, P. A., & Newbold, C. J. (2005). Rumen microbial population dynamics in response to photoperiod. *Letters in Applied Microbiology, 41*, 97-101.

Morgan, K. N., & Tromborg, C. T. (2007). Sources of stress in captivity. *Applied Animal Behaviour Science, 102*, 262-302.

Morita, T., & Tokura, H. (1998). The influence of different wavelengths of light on human biological rhythms. *Applied Human Science, 17*, 91-96.

Narins, P. M., Lewis, E. R., Jarvis, J. J. U. M., & O'Riain, J. (1997). The use of seismic signals by fossorial southern African mammals: A neuroethological gold mine. *Brain Research Bulletin, 44*, 641-646.

Norton, J. N., Kinard, W.L., & Reynolds, R. P. (2011). Comparative vibration levels perceived among species in a laboratory animals facility. *Journal of the American Association for Laboratory Animal Science, 50*, 653-659.

Nowacek, D. P., Thorne, L. H., Johnston, D. W., & Tyack, P. L. (2007). Responses of cetaceans to anthropogenic noise. *Mammal Review, 37*, 81-115.

O'Connel-Rodwell, C.E., Wood, J.D., Rodwell, T.C., Puria, S., Partan, S.R., Keefe, R., Shriver, D., Arnason, B.T., Hart, L.A. (2006). Wild elephant (*Loxodonta africana*) breeding herds respond to artificially transmitted seismic stimuli. *Behavioral Ecology and Sciobiology*, 59, 842-850.

OSHA, Occupational Safety and Health Administration. (1981). Occupational noise exposure. https://www.osha.gov/SLTC/noisehearingconservation/index.html (Accessed August 19, 2016).

Owen, M. A., Swaisgood, R. R., Czekala, N. M., Steinman, C. K., & Lindburg, D. G. (2004). Monitoring stress in captive giant pandas (*Ailuropoda melanoleuca*): Behavioral and hormonal responses to ambient noise. *Zoo Biology, 23*, 147-164.

Peplow, M. (2013). The anatomy of sleep: The ebb and flow of neurotransmitters switches our brains between sleep and wakefulness in carefully regulated cycles. *Nature, 497*, S2-S3.

Perremans, S., Randall, J. M., Rombouts, G., Decuypere, E., & Geers, R. (2001). Effect of whole-body vibration in the vertical axis on cortisol and adrenocorticotropic hormone levels in piglets. *Journal of Animal Science, 79*, 975-981.

Perret, M. (1997). Change in photoperiodic cycle affects life span in a prosimian primate (*Microcebus murinus*). *Journal of Biological Rhythms, 12*, 136-145.

Perret, M., & Schilling, A. (1995). Sexual responses to urinary chemosignals depend on photoperiod in a male primate. *Physiology and Behavior, 58,* 633-639.

Powell, D. M., Carlstead, K., Tarou, L. R., Brown, J. L., & Monfort, S. L. (2006). Effects of construction noise on behavior and cortisol levels in a pair of captive giant pandas (*Ailuropoda melanoleuca*). *Zoo Biology,* 25, 391-408.

Prescott, N. B., & Wathes, C. M. (1999). Spectral sensitivity of the domestic fowl (*Gallus g. domesticus*). *British Poultry Science, 40,* 332-339.

Prescott, N. B., Wathes, C. M., & Jarvis, J. R. (2003). Light, vision and the welfare of birds. *Animal Welfare, 12,* 269-288.

Prior, H. (2006). Effects of the acoustic environment on learning in rats. *Physiology and Behavior, 87,* 162-165.

Read, J., Jones, G., & Radford, A. N. (2014). Fitness costs as well as benefits are important when considering responses to anthropogenic noise. *Behavioral Ecology, 25,* 4-7.

Ross, M. R., Gillespie, K. L., Hopper, L. M., Bloomsmith, M. A., & Maple, T. L. (2013). Differential preference for ultraviolet light among captive birds from three ecological habitats. *Applied Animal Behaviour Science, 147,* 278-285.

Shannon, G., McKenna, M. F., Angeloni, L. M., Crooks, K. R., Fristrup, K. M., Brown, E., . . . Wittemyer, G. (2015). A synthesis of two decades of research documenting the effects of noise on wildlife. *Biological Reviews, 91,* 982-1005. doi:10.1111/brv.12207

Shepherdson, D. J., Carlstead, K. C., & Wielebnowski, N. (2004). Cross-institutional assessment of stress responses in zoo animals using longitudinal monitoring of faecal corticoids and behaviour. *Animal Welfare, 13,* S105-113.

Šklíba, J., Šumbera, R., & Chitaukali, W. N. (2008). Reactions to disturbances in the context of antipredatory behaviour in a solitary subterranean rodent. *Journal of Ethology, 26,* 249-254.

Smith, M. G., Croy, I., Ögren, M., & Waye, K. P. (2013). On the influence of freight trains on humans: A laboratory investigation of the impact of nocturnal low frequency vibration and noise on sleep and heart rate. *PLoS ONE, 8,* e55829. doi: 10.1371/journal.pone.0055829

Smith, M. G., Croy, I., Hammar, O., & Waye, K. P. (2016). Vibration from freight trains fragments sleep: A polysomnographic study. *Scientific Reports, 6,* 24717; doi: 10.1038/srep24717-

Sobrian, S. K., Vaughn, V. T., Ashe, W. K., Markovic, B., Djuric, V., & Jankovic, B. D. (1997). Gestational exposure to loud noise alters the development and postnatal responsiveness of humoral and cellular components of the immune system in offspring. *Environmental Research, 73,* 227-241.

Soltis, J. (2010). Vocal communication in African elephants (*Loxodonta africana*). *Zoo Biology, 29,* 192-209.

Stevens, R. G. (2006). Artificial lighting in the industrialized world: Circadian disruption and breast cancer. *Cancer Causes & Control, 17,* 501-507.

Stoinski, T. S., Jaicks, H. F., & Drayton, L. A. (2012). Visitor effects on the behavior of captive western lowland gorillas: The importance of individual differences in examining welfare. *Zoo Biology, 31*, 586-599.

Stone, E. L., Jones, G., & Harris, S. (2009). Street lighting disturbs commuting bats. *Current Biology, 19*, 1123-1127.

Van der Meer, E., Van Loo, P. L. P., & Baumans, V. (2004). Short-term effects of a disturbed light-dark cycle and environmental enrichment on aggression and stress-related parameters in male mice. *Laboratory Animals, 38*, 376-383.

Videan, E. N., Fritz, J., Howell, S., & Murphy, J. (2007). Effects of two types and two genre of music on social behavior in captive chimpanzees (*Pan troglodytes*). *Journal of the American Association for Laboratory Animal Science, 46*, 66-70.

Wale, M. A., Simpson, S. D., & Radford, A. N. (2013). Noise negatively affects foraging and antipredator behaviour in shore crabs. *Animal Behaviour, 86*, 111-118.

Walton, J. C., Weil, Z. M., & Nelson, R. J. (2011). Influence of photoperiod on hormones, behavior, and immune function. *Frontiers in Neuroendocrinology, 32*, 303-319.

Weil, Z. M., Bowers, S. L., & Nelson, R. J. (2007). Photoperiod alters affective responses in collared lemmings. *Behavioural Brain Research, 179*, 305-309.

Wells, D. L., & Irwin, R. M. (2008). Auditory stimulation as enrichment for zoo-housed Asian elephants (*Elephus maximus*). *Animal Welfare, 17*, 335-340.

Westlund, K., Fernström, A. L., Wergård, E. M., Fredlund, H., Hau, J., & Spångberg, M. (2012). Physiological and behavioural stress responses in cynomolgus macaques (*Macaca fascicularis*) to noise associated with construction work. *Laboratory Animals, 46,* 51-58.

Yokoyama, S., Takenaka, N., Agnew D.W., Shoshani, J. (2005). Elephants and human color-blind deuteranopes have identical sets of visual pigments. Genetics, 170, 335-344.

Zhao, Z., & Wang, D. (2005). Short photoperiod enhances thermogenic capacity in Brandt's voles. *Physiology and Behavior, 85,* 143-149.

Chapter 9

FEARLESS, PARTY OF TWO, YOUR TABLE IS READY!

Acknowledging Fears and Finding Commonalities to Facilitate Collaboration for Improved Elephant Well-Being

Margaret Whittaker
Creative Animal Behavior Solutions

F ear alters behavior. We talk about fear altering an animal's behavior and the inability to escape fear leading to a state of distress. Animals may have a difficult time learning, socializing, and problem solving when continuously exposed to fear-provoking stimuli. Fortunately, we care deeply about our charges and employ techniques to help them overcome fear and learn to tolerate, or even enjoy, that which was previously frightening. We do this because . . . dare I say . . . because we love these individuals and want to provide them with the best possible life in captivity.

Human behavior is not so different from animal behavior, including being motivated by fear. This paper is most likely different from others included in this book both in content and purpose; it is intended to generate kind, provocative discussion; thoughtful debate; and even conflict, which I hope will offer the opportunity for resolution. Facilities caring for elephants evolved from very different perspectives, but most claim to provide the best care for the elephants. The ideal way to care and provide for captive elephants continues to be a contentious topic worldwide. Elephants, with their numerous reasons for captivity (work, entertainment, exhibitory, education, religion, and sanctuary) and the resulting and divergent management styles, continue to garner considerable attention from professional animal care communities, concerned private citizens, private elephant owners, and legislators. Elephant management is the topic of considerable antagonistic debate, which has yielded an environment of assumptions, noncollaboration, and mistrust. This chapter will include a historical perspective with the intention to propose why we are functioning in this zone of fear and mistrust, recognize differences in elephant management and purpose, and recommend finding common ground which can begin to open the door for improved communication and, ultimately, elephant wellness.

I'd like to suggest that the professional animal care community are moved to action or, conversely, remain content with inaction too often because of fear. However, there is good reason to be fearful; history and experience have made us suspicious and mistrusting, and we consider those working in different types of facilities "on the other side." No doubt we have just cause for this mistrust: often our words have been taken out of context, we've felt betrayed by a trusted colleague, or we've been unfairly criticized for something beyond our control. History cannot and should not be ignored; rather, we acknowledge the highly opinionated and contentious environment that has surrounded elephant management, driven decision-making, divided the industry, and diminished collaboration, which may have yielded a negative effect on elephant care and well-being. Elephants continue to be the subject of unprecedented attention and scrutiny; therefore, open discussion is essential as we continue along the path to improving conditions for elephants in captivity.

I have learned so much over the years from animals in drastically different living conditions. Those individuals who were abused, mistreated, neglected, or somehow harmed by people are among the most remarkable of teachers. I have observed them as they recover their dignity, rediscover autonomy, and realize their independence. The behavioral flexibility demonstrated by these individuals never ceases to amaze me. Their ability to forgive, learn, and persevere is truly remarkable. For them, we must keep talking to one another, overcome the fear, and address the mistrust.

My time working with people and animals in zoos, aquariums, safari parks, biomedical facilities, and sanctuaries has given me a strong sense of likeness and similarity rather than one of differences. This chapter will focus primarily on two types of facilities housing elephants: zoological and sanctuary. I will propose a path of collaboration rather than division. The practice of keeping elephants in zoos or menageries has existed for hundreds of years. Comparatively, animal sanctuaries are relatively new and have existed for less than 50 years (Global Federation of Animal Sanctuaries, 2018b). Elephant sanctuaries have been in existence since the 1980s (Performing Animal Welfare Society, 2018). Zoos evolved from menageries where animals were kept as oddities or strange creatures for human entertainment and use. Sanctuaries grew from the need to provide for animals neglected and mistreated by humans. While it's true these facilities evolved from drastically different beginnings and maintain animals for strikingly dissimilar reasons, a comparative look at mission statements and guiding principles from zoos and sanctuaries may reveal more commonalities than many would suspect. We could choose to focus on similarities but historically have chosen to build walls and emphasize differences, resulting in an atmosphere of mistrust and fear.

SAMPLE MISSION STATEMENTS AND GUIDING PRINCIPLES FROM ZOOS AND SANCTUARIES

Facility 1 – Connecting communities with animals, inspiring action to save wildlife

- Be a leader in the global movement to save wildlife.

- Provide exemplary animal care, assuring outstanding animal welfare.

- Provide engaging animal experiences.

- Be a leading environmental education resource.

- Inspire broad community support and collaborations.

- Change behaviors to protect wildlife.

- Operate sustainably; be an example for others.

Facility 2 – Provide home, refuge, and individualized care for life

- Provide captive animals with individualized care, companionship, and the opportunity to live out their lives in a safe haven dedicated to their well-being.

- Educate the public of the complex needs of animals in captivity and the crisis they face in the wild.

Facility 3 – Committed to saving species worldwide by uniting our expertise in animal care and conservation science with our dedication to inspiring passion for nature

Facility 4

- Demonstrate leadership in wildlife conservation and animal welfare.

- Provide a broad audience with outstanding and unique educational opportunities that lead to the appreciation and stewardship of nature.

- Inspire our community with engaging, meaningful, and memorable experiences.

- Provide innovative facilities that contribute to the region's economic vitality.

- Demonstrate organization excellence consistent with a commitment to outstanding service, progressive resource management, and environmental leadership.

Facility 5

- Dedicated to the protection of animals.

- Provide permanent home for abused, abandoned, and retired captive wildlife.

- Enforce best standards of care for all captive wildlife.

- Preservation of wild species and their habitats.

- Promote public education about captive wildlife issues.

Facility 6

- Provide and promote best care practices.

- Inspire action for wildlife worldwide.

- Offer lifetime quality care for formerly abused and exploited animals and advocate for the species.

Facility 7

- Offer animals a place to retire.

- Re-socialize and rehabilitate animals.

- Provide information to support researching animals' complex needs and behaviors.

- Promote respect and protection for animals and their habitats.

Mission statements summarize the "why" an organization exists and provide a path to create guiding principles upon which the "how" and "what" are implemented to achieve the mission. Even this cursory look at mission statements demonstrates conspicuous similarities such as best practices in animal care, animal welfare/well-being, protecting wild animals and habitats, inspiring respect for natural world, and promoting conservation education. Of these facilities, four are sanctuaries and three are zoos. Key words such as "retire," "rehabilitate," or "permanent home" are associated with sanctuaries, and similarly, "connecting people to wildlife" or "animal experiences" are more often associated with zoos.

As I look at these mission statements, I find more in common than different, even at the organization level. When considering animal care and daily operations, the differences between zoos and sanctuaries become even less obvious than at the institutional or organizational level. Yet it is at this level where the majority of disagreement has historically occurred among those providing care of elephants. This brings me back to the question: are zoos and sanctuaries really so different? I imagine that most would answer, "YES!!!" The immediate and almost kneejerk answer of "YES!" should make us stop and examine the facts. Of course, there are obvious differences in zoos and sanctuaries, including:

- Reproduction objectives – Zoos selectively and purposefully breed animals for genetic and demographic management. Sanctuaries purposefully do not breed.

- Length of residency – Zoos move animals among accredited facilities for breeding, socialization, and exhibit purposes. Sanctuaries typically provide lifelong housing unless relocation is to meet socialization or behavioral needs.

- Frequency of public visitation – Zoos are open to the public most of the time. Sanctuaries may have scheduled "open houses" or special donor events.

- Genesis of institutions

- *Perception* of purpose

Keeping in mind the nature of the specific differences, consider the scope of the similarities:

- Animal welfare agenda

- Provide optimal husbandry

- Excellent veterinary care

- Application of enrichment strategies that are needs-driven and purposeful

- Limited resources

- Retro-fit behavioral management into existing care system

- Rigorous accreditation process for AZA and GFAS

- Standard guidelines for care for both AZA- and GFAS-accredited facilities

In consideration of these similarities and differences, I'm left wondering if we shouldn't be asking the question: are there enough similarities that we can collaborate for the betterment of animal care and well-being? It is this question I'd like to explore with an open mind, kind and respectful discussion, and an examination of how collaboration can be mutually beneficial if we can garner the strength to overcome our fears and mistrust.

Working in both zoos and sanctuaries for over 30 years has given me a unique perspective and has, on occasion, left me feeling like a tennis ball at Wimbledon—furiously bouncing back and forth between players. Fortunately, there have been sage visionaries on each side of the court who have attempted to forge relationships and break down barriers. These visionaries have entrusted me to work with both staff and animals, allowed me to share techniques and experiences, and encouraged me to open my eyes to the benefits of collaboration. Sanctuaries (housing all types of animals) and zoos have collaborated to improve conditions for individual animals, broadened the knowledge base, and supported rescue efforts. This long-standing history of working together cannot be overlooked and I hope will inspire more of us to seek common ground.

Both zoos and sanctuaries have come a long way in recognizing the need for rigorous accreditation processes, guiding principles that give direction across the profession, and continuing to raise the standards for accredited facilities. Both AZA (2018 and GFAS (2018a) accreditation processes include similar elements: care standards, housing requirements, veterinary care, staff levels and training, safety protocols/policies/training, well-being standards, financial stability and planning, governing authority, educational messaging, and public interaction policies, to name a few. These parallel accreditation processes reveal continued change and a commitment to excellence. They strengthen the argument for collaboration while respecting differences. Collaboration does not require sameness or compromise, but instead is founded in the idea that all parties recognize the best solution comes from working together as partners. Elephant management has changed considerably over the past decades with emphasis on enhanced environments and larger spaces, recognition of the importance of larger social groups, improved veterinary diagnostics and treatment, and a steady move in the direction of positive reinforcement training since the advent of protected contact management. Most sanctuary and zoo elephant care professionals recognize the essential need for a comprehensive management and care system. To clarify, a "system" of management necessitates a definable method including tools, techniques, and underlying tenets which can be applied in various situations and contexts (Whittaker & Laule, 2009). It must be able to meet all care needs and be transferable among all staff members, both within a single facility and among institutions. In other words, phrases such as, "only X person can do that" (assuming proper training has taken place) aren't consistent with this definition of a "system." A consistent approach yields this type of system. Consistency does not equal absolute sameness, and as a result, elephant management at any given facility will likely vary slightly due to accommodating the individual elephants' needs, facility design differences, social or herd management, etc. Still, management strategies should meet the described criteria of a "system."

Zoos and sanctuaries face different challenges in elephant care. Nevertheless, both facilities must also address many of the same ones. From the perspective of daily care and management, both use behavioral management practices and have developed comprehensive behavioral management programs. For the purposes of this paper, behavioral management is defined as a comprehensive, proactive approach to managing animal behavior that relies on two technical elements—environmental enrichment and positive reinforcement training—and is supported by facility design and operations. Measurement is essential. Recording what is done and what impact these actions have on animals contributes to both a broader scientific understanding and helps guide daily decision-making (Desmond, 1994; Coleman, 2012; Olsen, 2005; Whittaker, 2007). Although behavioral management programs were first used by zoos, they are now common practice in sanctuaries.

Zoos and sanctuaries arrived at the same point of using behavioral management from different perspectives, but both with similar purpose—the need to provide better care facilitated by improved access to animals. The intensive management typically used in zoos in addition to the need to improve implementation methods lead to the development of behavioral management (see Desmond, 1994; Olsen, 2005; Laule, 1993; Perlman et al., 2012). Challenges encountered by zoos commonly include situations exacerbated by limited space or holding options (e.g., social management, obesity, and related conditions). Sanctuaries initially held to the idea that large spaces in natural settings would return elephants to a state of physical health and fitness and mental well-being. While it's logical and demonstrable that these environments provide a vast improvement to the traditional manner in which elephants were maintained, a long life in inappropriate circumstances can result in conditions that require intervention and, therefore, the application of behavioral management. Challenges encountered by sanctuaries commonly include: rehabilitation and socialization, managing and accessing animals in large spaces, and creating plans to accommodate special-needs individuals. For both zoos and sanctuaries, resolutions to these challenges can be found when comprehensive, well-planned, goal-oriented behavioral management techniques are applied.

In a well-managed facility, daily care and husbandry for elephants, regardless of location, may look very much the same. Elephants are provided high-quality feed; enclosures are cleaned and checked for safety; husbandry care such as footwork, skin care, exercise, and environmental enrichment are assessed and/or provided; and veterinary care is provided as needed. Elephants are accessible either in barns, habitats, or other areas properly suited for training. The daily care administered to elephants can appear quite similar in facilities that use positive reinforcement and protected contact techniques. In both facilities, elephants receive foot care. In both facilities, elephants are treated for illnesses or injuries. In both facilities, elephants are trained for veterinary procedures such as blood collection, injections, radiographs and ultrasounds, rectal cleanouts, pill swallowing, laser treatments, or intensive treatments of foot maladies. With improvements to the conditions in which captive elephants live, we should see a decline of certain conditions, such as degenerative joint disease, and begin to realize the goal of healthier elephants.

Concentrating energies on collaboration to facilitate further advancements and improvements for elephants in human care is a big step. A past clouded with mistrust, anger, and fear cannot be ignored or forgotten, but if a few elephant care professionals follow those sage visionary's' leadership and begin to find common ground, admit fears, and focus on the future of elephant care, perhaps we can look to the upcoming years of elephant care through a very different lens.

REFERENCES

Association of Zoos and Aquariums (AZA). (2018). *Accreditation Standards and Related Policies, 2019 Edition.* Retrieved from https://www.aza.org/assets/2332/aza-accreditation-standards.pdf.

Coleman, K. (2012). Individual differences in temperament and behavioral management practices for nonhuman primates. *Applied Animal Behavior Science, 137*(3), 106–113.

Desmond, T. (1994). Behavioral management—An integrated approach to animal care. *Annual Proceedings of the American Zoological and Aquarium Association* (pp. 19–22).

Global Federation of Animal Sanctuaries (GFAS). (2018a). *Standards of excellence.* Retrieved from https://www.sanctuaryfederation.org/for-sanctuaries-2/standards/.

Global Federation of Animal Sanctuaries (GFAS). (2018b). https://www.sanctuaryfederation.org/.

Laule, G. (1993). The use of behavioral management techniques to reduce or eliminate abnormal behavior. *Animal Welfare Information Center Newsletter (USA).*

Olsen, Diane. (2005). Creating an enrichment and training program from scratch. Proceedings from the *AZAConference.*

Performing Animal Welfare Society. (2018). http://www.pawsweb.org/.

Perlman, J. E., Bloomsmith, M. A., Whittaker, M. A., McMillan, J. L., Minier, D. E., & McCowan, B. (2012). Implementing positive reinforcement animal training programs at primate laboratories. *Applied Animal Behavior Science, 137*(3), 114–126.

Whittaker, M. (2007). *Behavioral management and sanctuaries: A perfect fit for enhancing animal well-being.* Paper presented at the AAZK Annual Conference. Moody Gardens, Galveston, TX.

Whittaker, M., & Laule, G. (2009). Protected contact and elephant welfare. In D. L. Forthman, L. F. Kane, D. Hancocks, & P. F. Waldau (Eds.), *An elephant in the room: The science and well-being of elephants in captivity* (pp.181–188. North Grafton, MA: Tufts University Press.

Chapter 10

CRUX MOVES

Learning Solutions at the Zoo

S. G. Friedman

Professor Emeritus, Department of Psychology, USU

Behavior Works, LLC

Interest in positive reinforcement training strategies has gone from a ripple to a tsunami over the past two decades. This wave reached a *Scientific American* journalist who investigated a recent study that compared two approaches to training orthopedic surgical residents. Ironically, residents taught complex surgical skills with "dog training techniques" (a bridging stimulus, back-up social reinforcers, and shaping by approximations) outperformed those taught with the traditional approach (i.e., modeling, imitation, and social punishers: Konkel, 2016). The reporter contacted me, an applied behavior analyst who has worked with animal trainers and behavior consultants for nearly 20 years, to learn the answer to her last, most elusive question: why does it work?

Three "crux moves," or perspectives, from the science underlying training technology, are presented here to help answer Konkel's question and enrich professionals' understanding of the learning solutions they use to increase elephant wellness at zoos.

CRUX MOVE #1: THE STUDY OF BEHAVIOR CHANGE IS A LIFE SCIENCE

Konkel's question was interesting but not surprising. I have observed frequently that while people are familiar with the usual cast of evolutionary adaptations we use to interact with the environment (eyes to see, ears to hear, legs to run, etc.), learning is rarely on their list of evolved, biological mechanisms; however, it should be. As Chance (2009) stated, "Learning does not give the species the tendency to behave a certain way in a particular situation; rather it gives the individual the tendency to modify its behavior to suit a situation." Learning is our nature.

Behavior analysis is the natural science of learning. Learning is defined as behavior change due to experience (i.e., contact with certain environmental events). The goal of behavior analysis is the same as that in any natural science, which is to explain phenomena by identifying the *physical event*s that produce them. Mental constructs (e.g., psyche, impulse control, boredom) do not make good outcome measures. Constructs are unobservable events, the existence of which is inferred from observable behavior. As concepts, constructs have no tangible form and, therefore, cannot cause anything. In behavior analysis, the primary focus is on the observable environmental determinants of behavior.

There is more than one science of behavior (e.g., ethology, behavioral neuroscience, cognitive psychology). Each asks different questions and employs different methods to obtain answers. Contests with respect to which science is preeminent should be replaced with the more contemporary view that each investigates a different level of analysis. It will take more than one science to complete the learning and behavior puzzle; however, no account of learning is complete without the behavioral level of analysis (behavior-environment relations).

One of the benefits of a natural science approach to learning is the process of "self-correction" that results from the replication of findings and peer review. This process by which scientific knowledge is revised is often taken as proof that research results are as capricious as personal opinions or conventional wisdom. The difference is that scientific revisions are based on data that result from the application of the scientific method and not the "cultural fog" (an evocative term I borrowed from Swedish economist Gunnar Myrdal [as cited in Gould, 1981]).

The Study of One

One distinguishing characteristic of behavior analysis is its focus on the individual learner. This focus is particularly relevant to elephant wellness, as wellness is, ultimately, a study of one. Learning solutions are based on the general principles of behavior change and then are custom fit to each animal and context. In contrast, ethology focuses on typical behavior patterns of animal groups (species) in their natural environment. Historically, ethology has informed the management of captive animals living in unnatural environments in significant ways, but it is not sufficient. By adding the training dimension to the standard of care in zoos, animals can be taught behavioral repertoires that improve their quality of life in their current conditions. This includes those individuals not represented well by the species-level norms of their wild counterparts. Animals are biologically prepared to learn (i.e., to produce meaningful outcomes with their behavior) in both the wild and in captivity.

CRUX MOVE #2: THE ABCS OF BEHAVIOR

The smallest meaningful unit of behavior analysis consists of *antecedents, behavior,* and *consequences*. Talking about behavior in isolation from its antecedents and consequences just does not make sense. Behavior never occurs in a vacuum. Behavior is always conditional (i.e., dependent on the conditions in which it occurs). Behavior is what an animal *does* (verbs) that can be observed. Sparring with a log, shifting from one area to another, and presenting an ear for a blood draw are all behaviors. Descriptions such as dominant, intelligent, and jealous are adjectives that label animals too ambiguously and indirectly to proceed with specific training targets. Even the innocuous label, "friendly," needs to be defined operationally (e.g., approaches humans, stands relaxed while touched, takes food from hands)—these are the behaviors we train, and once the animal masters them, we call him/her friendly. When evaluating behavior, "unlabel" your learner.

The word *consequence* often connotes retaliation and retribution ("You'll suffer the consequences!"). However, in the learning context, consequences are the *outcomes* of behaving, the *purpose* for behaving, and *feedback* about the adequacy of behavior. To answer questions about why an animal behaves a certain way, we must ask three key questions: 1) what does the behavior look like in unambiguous, observable terms; 2) when is the behavior most likely to occur (antecedent conditions); and 3) what is its function (immediate consequences)? Identifying the behavior ABCs is the first step in understanding, predicting, and changing behavior systematically.

Skinner (1981) called the reciprocal relation between behavior and consequences the second kind of selection by consequences. Indeed, consequences provide the environmental pressure that selects or deselects an individual's behavior within its lifetime. Training (teaching) is, in part, the process of arranging consequences deliberately to select (reinforce) or deselect (extinguish) another individual's behavior. We can use these natural forces to benefit animals without becoming mired in philosophical discussions of free will. Gravity controls our behavior daily, yet we let Newton off the hook.

With respect to operant learning, antecedent events signal but do not cause behavior. Antecedent stimuli in the form of cues (i.e., discriminative stimuli), setting events, and motivating operations predict the behavior-consequence contingency ahead. For example, when you are driving, a red traffic light does not *cause* your foot to move to the brake; rather, it signals the opportunity for reinforcement (i.e., the car stopping in the face of oncoming traffic), which is *contingent on* putting your foot on the brake. Between antecedent events and behavior is *choice,* which is informed by the individual's learning history. No animal behaves to be ineffective.

Emotions

From the perspective of behavior analysis, the word *emotion* describes overt, covert, innate, and learned behaviors. As with other labels, we must construct careful operational definitions of emotional behavior. What do happy, afraid, anxious, and frustrated look like for a given species and, most importantly, this individual? Every behavior we train is associated with an emotion (Laying, 2017). Happy is a function of positive reinforcement contingencies, fear of escape contingencies, anxiety of avoidance contingencies, and frustration of extinction contingencies. Emotions are both the result of motivating operations and are motivating operations themselves—those antecedent events that change the strength of a reinforcer temporarily (Lewon & Hayes, 2014). As such, emotions should be included as a part of every comprehensive training program.

Neuroscience

I concede that neuroscience is the sexy science sister. It is undeniably exciting that new imaging technology can reveal the inner workings of live, healthy brains. However, this is a field in its infancy, and many recent examples of neuroscience findings include over-interpretation or misinterpretation of correlational studies. Epstein (2008) wrote: "The problem with many headlines these days is that they automatically claim, based on the latest correlational brain study, that we have identified the cause of depression or love or autism or Alzheimer's just because some area of the brain lights up when people have that condition. But finding correlations isn't the same as finding causes, and finding causes is often quite difficult."

Cause is rarely a single agent. There are many causes because of all the different levels of analysis involved in a phenomenon. As with overt behavior, the behaving brain also depends on environmental conditions. The tools that change overt behavior by changing observable conditions should be at the top of the trainer's toolbox.

CRUX MOVE #3:
CONTROL IS A PRIMARY REINFORCER

In behavior analysis, consequences can be categorized loosely as innate (primary) or learned (secondary). Primary reinforcers include a relatively short list of consequences essential for survival (e.g., food, water, warmth). The list of potential secondary reinforcers (which result when a neutral stimulus is paired repeatedly with an existing reinforcer) is endless. Primary reinforcers do not require a conditioning history, while, with secondary reinforcers, there is always a reason to behave, even when your basic needs are met. Primary and secondary reinforcers are our best sources of motivation.

Research suggests that control is a primary reinforcer, which makes sense from an evolutionary point of view. To the greatest extent possible, we should provide animals in human care with control over significant life events. We can accomplish this by increasing the opportunity for them to make choices and solve problems in enriched environments. Three independent lines of scientific inquiry support this assertion.

Control

Watson (1967, 1971) compared two groups of 90-day-old babies who had moving mobiles above their cribs. The amount of time the mobiles moved was the same for both groups. However, only one group had the ability to control the onset of the motion by raising their heads to close a circuit hidden under their pillows. Initially, both groups responded to the moving mobiles by cooing and smiling, a reasonable measure of one important component of wellness. Those happy responses continued throughout the experiment among the babies who controlled their mobiles. However, for those who could not control their mobiles, the cooing and smiling stopped quickly. This evidence suggests that moving mobiles reinforce head-lifting behavior, but more impressively, the evidence also suggests that *controlling* the movement of the mobiles increases positive emotional responses.

Joffe, Rawson, and Mulick (1973) observed that 60-day-old rats provided with response contingent lighting, food, and water were more active, exploratory, and less emotional (reduced defecation) in a novel arena. Similarly, Mineka and Henderson (1985) found that Rhesus macaques reared with control over access to reinforcers (food, water, treats) were bolder engaging with novel toys and novel rooms and when separated from peers at six to 10 months old.

Contrafreeloading

Research has demonstrated that, given a choice between proffered food or food that requires behavioral effort, animals tend to prefer the food that requires effort. This tendency is called contrafreeloading (or the free food phenomenon) and has been replicated with rats, mice, chickens, pigeons, crows, cats, gerbils, Siamese fighting fish, humans (Osborne, 1977), starlings (Inglis & Ferguson, 1986), Abyssinian ground hornbills, bare-faced curassows (Gilbert-Norton, 2003), and captive parrots (Coulton, Warren, & Young, 1997).

Hypotheses about why animals prefer generally to work for food run the gamut (e.g., frustrated foraging motor patterns in captivity; information-seeking behavior to locate optimal food sources; additional reinforcement provided by stimulus changes when one works for food, such as the sound of a hopper). In any case, the free food phenomenon supports strongly the general hypothesis that behavioral control is important to animals.

Learned Helplessness

In contrast, investigations of learned helplessness have provided evidence that a lack of control can have pathological effects, including depression, learning disabilities, emotional problems (Maier & Seligman, 1976), and suppressed immune system activity (Laudenslager, Ryan, Drugan, Hyson, & Maier, 1983). Learned helplessness refers to the detrimental effects of blocking an animal's natural attempts to escape aversive stimulation. Subsequently, when escape *is* possible, animals continue to behave helplessly (i.e., remaining passively in the presence of the aversive stimulus). This research has been replicated with cockroaches (Brown, Hughes, & Jones, 1988), dogs, cats, monkeys, and human children and adults (Overmier & Seligman, 1967). Seligman (1990) suggested that learners can be "immunized" against learned helplessness by providing them with abundant experiences in which their behavior is effective. Thus, control as a lifestyle builds the resilience we need to withstand occasional, inevitable lack of control.

A Suggested Guideline

Given the research evidence that control is a primary reinforcer (i.e., a biological need), the effectiveness of training alone is an insufficient standard for providing and ensuring elephant wellness. Of course, training should be effective, but at what price to the animal? Not every behavior that can be trained should be trained; further, not just any effective procedure should be used. Carter and Wheeler (2005) defined *intrusiveness* as the degree to which a procedure affords the learner counter control. The characteristic of intrusiveness adds significantly to a trainer's ethical standards (for a more complete treatment of this topic, see Friedman, 2008). Figure 1 shows an expanded guideline (based on Alberto & Troutman, 1999) for selecting training procedures according to a hierarchy of intrusiveness. It is no small thing to change an animal's functional behavior. In all cases, the goal should be to use the least intrusive effective training strategy.

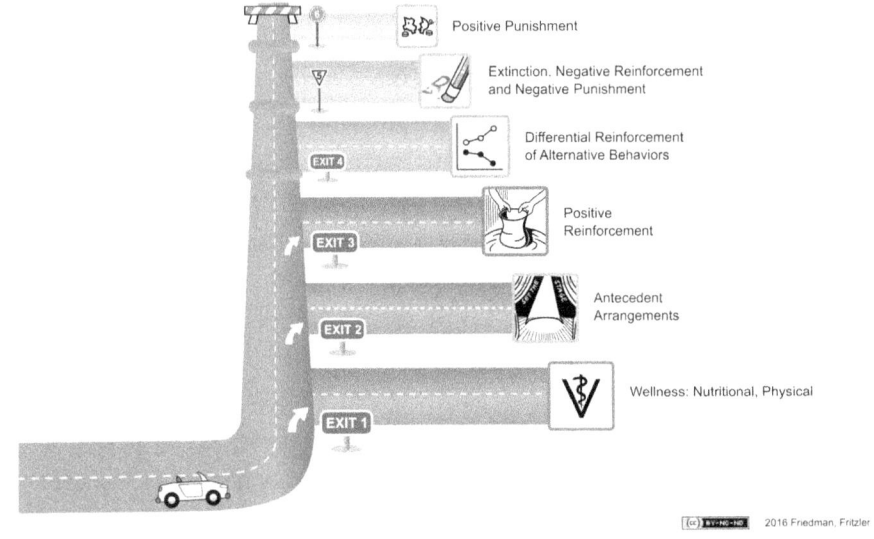

Positive Punishment

Extinction. Negative Reinforcement and Negative Punishment

Differential Reinforcement of Alternative Behaviors

Positive Reinforcement

Antecedent Arrangements

Wellness: Nutritional, Physical

EXIT 4

EXIT 3

EXIT 2

EXIT 1

2016 Friedman, Fritzler

FIGURE 1: HIERARCHY OF BEHAVIOR CHANGE PROCEDURES FROM LEAST TO MOST INTRUSIVE. "EXIT 1" IS THE LEAST INTRUSIVE INTERVENTION, AND "EXIT 6" IS THE MOST INTRUSIVE.

CONCLUSION

At the beginning of this paper, I shared the challenge of explaining to a reporter why "dog training techniques" worked with orthopedic residents learning complex surgical skills. Of course, we know that the techniques used in that study are applicable universally across species and not even remotely specific to *Canis familiaris*. I hope she understands that now too. The ability to change what we do based on experience is indeed a remarkable adaptation. Learning is required to survive the demands of an ever-changing environment, "an evolved modifiability" (Chance, 2009). As long as earth continues to be "change central," its animal life will continue to be "wicked good" learners.

Behavior does not spray out of animals like water from a leaky faucet. Behavior is lawful. Antecedent events give order to behavior by indicating which response is most likely to lead to reinforcement. When the doorbell rings, we open the door, not the microwave. Ultimately, behavior is a function of consequences. We would make very different predictions about the future frequency of door-opening if the consequences were court summonses and not boxes from Apple.

Behavior empowers animals to operate on the environment to protect and provide for themselves (and others). When we block control over significant life events continuously, animals are unable to thrive. They become helpless and lack the resilience to recover from life's occasional hardships. With thoughtful training, we change the conditions, and the animal changes its behavior.

According to Bailey and Burch (2005), what makes behavior analysis unique is relevant to professionals working with nonhuman animal behavior as well: the learners are often very vulnerable and unable to protect themselves from harm. This similarity suggests that the ethical standards established for behavior analysts also may have widespread relevance for professionals working with any species:

- Protect the participants' welfare at all times.

- Use interventions that are customized for each individual.

- Design interventions based on a functional assessment of the problem behavior.

- Use only procedures for which there is a scientific basis (evidence-based treatment).

- Use scientific methods to implement and evaluate interventions (e.g., collect pre-intervention baseline data and ongoing treatment data until the intervention is terminated).

With a sound knowledge of the fundamental principles of learning and their skillful application, we can be confident that learning solutions are, and will continue to be, a critical component of elephant wellness in zoos.

REFERENCES

Alberto, P. A., & Troutman, A. C. (1999). *Applied behavior analysis for teachers*. Upper Saddle River, NJ: Merrill Prentice Hall.

Brown, G. E., Hughes G. D., & Jones, A. A. (1988). Effects of shock controllability on subsequent aggressive and defensive behaviors in the cockroach (*Periplaneta americana*). *Psychological Reports, 63*, 563–569.

Bailey, J. S., & Burch, M. R. (2005.) *Ethics for behavior analysts*. Mahwah, NJ: LEA.

Carter, S. L., & Wheeler, J. J. (2005). Considering the intrusiveness of interventions. *The International Journal of Special Education, 20*, 132–142.

Chance, P. (2009.) *Learning and behavior.* Belmont, CA: Wadsworth Cengage Learning.

Coulton, L. E., Warren, N. K., & Young, R. J. (1997). Effects of foraging enrichment on the behavior of parrots. *Animal Welfare, 6*, 357–363.

Epstein, R. (2008, September/October). The truth about brain science. *Skeptical Inquirer, 32*–33.

Friedman, S. G. (2008). What's wrong with this picture? Effectiveness is not enough. *Good Bird Magazine, 4*(4), 12–18.

Gilbert-Norton, L. (2003). Captive birds and freeloading: The choice to work. *Research News, 4*(1).

Gould, S. J. (1981). *The mismeasure of man*. New York, NY: W.W. Norton.

Inglis, I. R., & Ferguson, N. J. K. (1986). Starlings search for food rather than eat freely available food. *Animal Behavior, 34,* 614–616.

Joffe, J., Rawson, R., & Mulick, J. (1973). Control of Their Environment Reduces Emotionality in Rats. *Science, 180*(4093), 1383-1384. Retrieved from http://www.jstor.org.dist.lib.usu.edu/stable/1736668

Konkel, Lindsey. (2016, March 9). Positive reinforcement helps surgeons learn. *Scientific American*. Retrieved from https://www.scientificamerican.com/article/positive-reinforcement-helps-surgeons-learn/.

Laudenslager, M. L., Ryan, S. M., Drugan, R. C., Hyson, R. L., & Maier, S. F. (1983). Coping and immunosupression: Inescapable but not escapable shock suppresses lymphocyte proliferation. *Science, 221,* 568–570.

Layng, T.V.J. (2017). Private emotions as contingency descriptors: emotions, emotional behavior, and their evolution. *European Journal of Behavior Analysis,* 18:2, 168-179.

Lewon, M., & Hayes, L. J. (2014). Toward an analysis of emotions as products of motivating operations. *Psychological Record, 64,* 813–825.

Maier, S. F., & Seligman, M. E. P. (1976). Learned helplessness: Theory and evidence. *Journal of Experimental Psychology: General, 105,* 3–46.

Osborne, S. R. (1977). The free food (contrafreeloading) phenomenon: A review and analysis. *Animal Learning & Behavior, 5*(8), 221–235.

Overmier, J. B., & Seligman, M. E. P. (1967). Effects of inescapable shock upon subsequent escape and avoidance responding. *Journal of Comparative and Physiological Psychology, 63*, 28–33.

Seligman, M. E. P. (1990). *Learned optimism.* New York, NY: Knopf.

Skinner, B. F. (1981). Selection by consequences. *Science, 213*, 501–504.

Watson, J. S. (1967). Memory and "contingency analysis" in infant learning. *Merrill-Palmer Quarterly, 13*, 55–76.

Watson, J. S. (1971). Cognitive-perceptual development in infancy: Setting for the seventies. *Merrill-Palmer Quarterly, 12*, 139–152.

Chapter 11

THE ELEPHANT IN THE ROOM

Kate E. Evans
Elephants for Africa

INTRODUCTION

Male African elephants (*Loxodonta africana*) have different ecological and social requirements than females (Stokke & du Toit, 2002; Evans & Harris, 2012). These often are overlooked in their management in both in-situ and ex-situ populations and have implications for their wellness and conservation.

In 2011, among the male African elephants recorded by the Association of Zoos & Aquariums (AZA) elephant TAG in North America, nearly two-thirds (63%) had no male companion (Olson, 2011). Of the 66 institutions involved in this survey, only one did not house female elephants (Birmingham Zoo, Alabama). Among those that housed males, eight had two or more male elephants (42%: refer to Figure 1), and two had three or more males (10%).

The AZA recommends that facilities should house at least three females, two males or more, or three elephants of mixed gender (Association of Zoos & Aquariums, 2012) while Roocroft and Zoll (1994) recommended a herd size of at least six females. The AZA standards state that male elephants of six years or older can be housed alone, although not in complete isolation, and must have opportunities for tactile, olfactory, visual, and auditory interaction with other elephants (AZA, 2012). These standards also state that new facilities must include a holding space for one adult male; there are no current standards for all male groups. Schulte

(2000) reported that of the 15 facilities at the time that housed more than one male elephant of either species, 10 allowed no physical interaction. Taylor and Poole's (1998) survey of European and North American zoos found that males always were kept separate from females.

FIELD STUDIES OF WILD AFRICAN MALE ELEPHANTS

The social life of a male elephant varies by age and can be divided into life stages that reflect different needs. Initially, they are dependent on their mothers and the herd into which they were born, but even here, behavioral differences between males and females manifest themselves (Lee & Moss, 2011). Their social interactions focus on male age mates while they are juveniles (Lee, Poole, Njiraini, Sayialel, & Moss, 2011), and when they leave their natal herd at adolescence, they spend time with other males, other breeding herds, or alone (Chiyo et al., 2011; Evans & Harris, 2008; Lee et al., 2011; Poole, 1982; Poole, Lee, Njiraini, & Moss, 2011). Wild males begin to experience their first musth in their 20s (Poole, 1987, 1989; Poole et al., 2011), which is characterized by physical and behavioral changes, with increased dominance displays and competitive interactions with other males, as well as increased sexual interest in females (Lee et al., 2011). During adulthood, their social and ecological requirements are influenced strongly by their musth periods, during which they seek out receptive females and chase competitors away and thereafter return to their feeding and bull areas (Poole, 1982; Moss & Poole, 1983).

What is clear is that male elephants socialize with other elephants, specifically other male elephants, throughout their lives, and males as young as five spend time away from their families (Lee et al., 2011). However, the average age at which a male leaves his herd and becomes independent is 14 (Douglas-Hamilton, 1972, 1973; Poole, 1987, 1994, 1996; Lee & Moss, 1999). Poole (as cited in Ogden, 2014) reported that

"Up until about 25 years they (male elephants) are spending 80% of their time in family groups."

In the wild, males spend time in areas inhabited primarily by bulls that are located away from where the majority of the female population resides (Evans, 2006; Lee et al., 2011). It appears that one primary purpose of a "bull area" is to serve as one in which mature bulls can gain weight and energy reserves before coming into musth and returning to predominantly female areas to find mating opportunities (Laws, Parker, & Johnstone, 1975; Moss & Poole, 1983; Moss, 1988). However, observations in Botswana showed that bulls of all ages use these areas (Evans, 2006; Evans, 2019) and that they may play an important social role in bull ecology. While we do not yet know why bulls reside in these different areas, we do know that age has a profound effect on their ecological and social requirements and that their individual personalities have an effect as well (Evans, 2006; Evans & Harris, 2008, 2012; Lee et al., 2011).

The mean group size of wild male elephants studied in the Okavango Delta, Botswana, was two, with all male groupings varying in size from one to 17 individuals (Evans & Harris, 2008). In the Makgadikgadi Pans National Park, groups of 100 or more male elephants have been seen regularly by the river (Evans, 2019) and other water sources (James Bradley, personal communication, June 2011).

The time spent in all-male groups varies with age and reproductive state (Chiyo et al., 2011; Evans, 2006), with younger males more often sighted in larger groups (Evans & Harris, 2008). While male elephants of all ages prefer to spend time with males of a similar age (Chiyo et al., 2011; Evans & Harris, 2008), they also prefer to associate with adult males above or equal to 36 years old (Evans & Harris, 2008). Thus, these groupings of male elephants are not random (Evans & Harris, 2008). Males have preferred companions, with older individuals forming stronger associations and younger individuals having more differentiated relationships (Elephants for Africa, unpublished data).

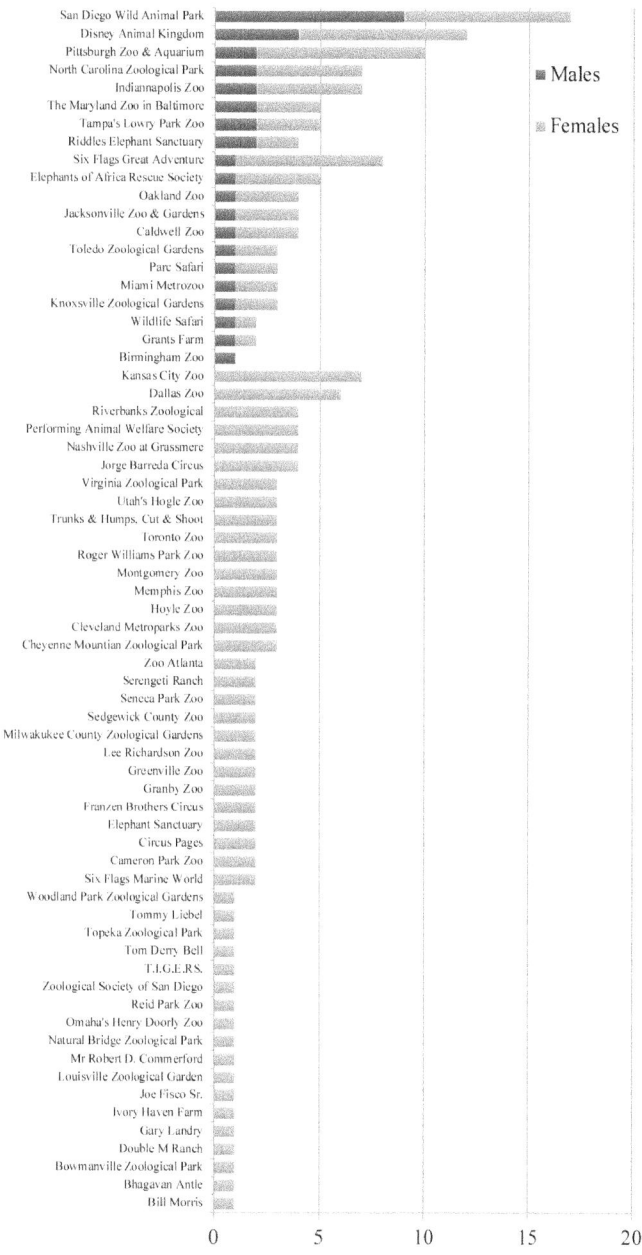

Total number of African elephants housed at location.

FIGURE 1: SOCIAL GROUPINGS OF AFRICAN ELEPHANTS IN NORTH
AMERICA AS SURVEYED BY THE AZA (OLSON, 2011).

The importance of older males in bull elephant society is often over-looked in the management of ex-situ populations through translocation programs, trophy hunting, and in in-situ populations where contact with older males is limited, if available at all.

In the past, some founder populations in South Africa were composed of the orphans of culls, with devastating consequences. In one national park alone, young bulls that had come into musth (Hanks & McIntosh, 1973; Laws, 1969; Lee, 1986; Poole, 1982, 1994; Short, Mann, & Hay, 1967) killed more than 40 white rhino (*Ceratotherium simum*) in a five-year period (Slotow, Van Dyk, Poole, Page, & Klocke, 2000; Slotow & Van Dyk, 2001; Slotow, Balfour, & Howison, 2001). This unusual behavior ceased when older males were introduced into the area and the younger males came out of musth. This and similar incidents in other parks and in private ownership (Evans, 2006) highlight the important role older males play in regulating bull society.

It also has become evident that the management of elephant populations is not only about numbers, but also about the dynamics of the population and the individuals within it. Therefore, young bulls, which often do not have contact with older bulls in captivity, are often separated prematurely from their mothers and housed alone in zoos (Clubb & Mason, 2002; Schulte, 2000; Taylor & Poole, 1998). Clearly, this is not conducive to the normal behavior observed in the wild.

As well as affecting their social groupings and interactions with other elephants, age affects a male elephants' behavior with adolescent males. The age group of 16- to 20-year-old males engages in higher rates of social activities, including sparring and greeting, the latter significantly more (Evans, 2006). This is associated with the period during which most male elephants have left their herds and begun their journeys as independent males. Thus, in the wild, they are meeting new males and integrating into male society.

The process of becoming independent can take between five months and eight years (Lee et al., 2011) and can be very individual, with some males leaving their herd and joining another family for a while. Others join groups of bulls, while yet others may remain in their natal herd's home range and move from family to family (Poole, 1996).

Adult males often are referred to as solitary. However, the median time males spent alone during sexually inactive periods in Amboseli National Park, Kenya, was only 9.8%, while in Botswana, adult males were seen alone most frequently (Evans, 2006), with individual differences having an effect. However, both of these studies are limited by the way in which the researchers perceive the world and cannot take into consideration the elephants' perception. For example, an apparently lone male elephant may be communicating with another through infrasound and out of visual contact.

APPLYING THE FINDINGS OF FIELD STUDIES TO ENHANCE MALE ELEPHANTS' WELLNESS IN ZOOS

Returning to the way male elephants are managed in zoos discussed in the introduction, how can we apply the scientific findings of field studies to enhance the existing standards and situations?

A review of the current situation needs to be urgently undertaken with clear motions and agreements in place to change current practices of keeping male elephants separated from other male elephants and lacking social contact with other elephants.

Elephants' ability to communicate over vast distances using infrasound, together with their exceptional sense of smell, should be considered in management decisions. When considering elephants in captivity, the human visitors' needs generally are the priority. Knowledge of the way in which elephants perceive their world could be incorporated into the development and design of their living quarters and could improve their wellness significantly. The sounds and smells of a busy human environment may be stressors to an animal whose sense of smell and sound are so acute.

In conclusion, male elephants are highly social, and their social requirements vary with age. However, males of all ages need contact with other elephants, with other males, and specifically with older males. The way elephants are managed and housed in captivity currently does not address their social requirements or, indeed, their ecological requirements. If elephant wellness is to advance beyond merely coping and surviving in captivity and achieve the ideal of thriving, then their social and ecological requirements need to be altered drastically to accommodate what they experience in the wild in individual institutions and the long-term management policies of the meta population. What must be asked is, is this at all possible, and if so, how?

REFERENCES

Association of Zoos & Aquariums (AZA). (2012). *Standards for elephant management and care.* Retrieved from https://www.aza.org/ assets/2332/aza_standards_for_elephant_management_and_ care.pdf.

Chiyo, P. I., Archie, E. A., Hollister-Smith, J. A., Lee, P. C., Poole, J. H., Moss, C. J., & Alberts, S. C. (2011). Association patterns of African elephants in all-male groups: The role of age and genetic relatedness. *Animal Behavior, 81,* 1,093–1,099.

Clubb, R., & Mason, G. (2002). *A review of the welfare of zoo elephants in Europe* (A Report Commissioned by the RSPCA, University of Oxford, United Kingdom).

Douglas-Hamilton, I. (1972). *On the ecology and behavior of the African elephant: Elephants of Lake Manyara* (Doctoral dissertation). United Kingdom: University of Oxford.

Douglas-Hamilton, I. (1973). On the ecology and behavior of the Lake Manyara elephants. *East African Wildlife Journal, 14,* 1–403.

Evans, K. E. (2006). *The behavioral ecology and movements of adolescent male African elephant* (Loxodonta africana) *in the Okavango Delta, Botswana* (Doctoral dissertation). United Kingdom: University of Bristol.

Evans, K. E. (2019). Bull elephants – their importance as individuals in elephant societies. Africa Geographic Blog https:// africageographic.com/blog/bull-elephants-their-importance-as-individuals-in-elephant-societies/

Evans, K. E., & Harris, S. (2008). Adolescence in male African elephants, *Loxodonta africana*, and the importance of sociality. *Animal Behavior, 76*(3), 779–787.

Evans, K. E., & Harris, S. (2012). Sex differences in habitat use by African elephants (*Loxodonta africana)* in the Okavango Delta, Botswana: Is size really the deciding factor? *African Journal of Ecology, 50,* 277–284.

Hanks, J., & McIntosh, J. E. A. (1973). Population dynamics of the African elephant (*Loxodonta africana*). *Journal of Zoology, 169,* 29–38.

Laws, R. M., Parker, I. S. C., & Johnstone, R. C. B. (1975). *Elephants and their habitats: The ecology of elephants in North Bunyoro, Uganda.* Oxford, United Kingdom: Clarendon Press.

Lee, P. C. (1986). Early social development among African elephant calves. *National Geographic Research, 2,* 388–401.

Lee, P. C., & Moss, C. J. (1999). The social context for learning and behavioral development among wild African elephants. *Symposium for the Zoological Society, London, 62,* 102–125.

Lee, P. C., & Moss, C. J. (2011). Calf development and maternal rearing strategies. In C. J. Moss, C. Harvey, & P. C. Lee (Eds.), *The Amboseli elephants: A long-term perspective on a long-lived mammal* (pp. 224–237). IL: University of Chicago Press.

Lee, P. C., Poole, J. H., Njiraini, N., Sayialel, C. N., & Moss, C. J. (2011). Dynamics: Independence and beyond. In C. J. Moss, C. Harvey, & P. C. Lee (Eds.), *The Amboseli elephants: A long-term perspective on a long-lived mammal* (pp. 224–237). IL: University of Chicago Press.

Moss, C. (1988). *Elephant memories*. IL: University of Chicago Press.

Moss, C. J., & Poole, J. H. (1983). Relationships and social structure of African elephants. In R. A. Hinde (Ed.), *Primate social relationships: An integrated approach* (pp. 315–325). Oxford, United Kingdom: Blackwell.

Ogden, L.E. (2014, October 31). Male elephants are not the loners we once thought. BBC Earth. Retrieved from http://www.bbc.co.uk/earth/story/20141101-male-elephants-have-a-sweet-side.

Olson, D. (2011). *North American region African studbook for the African elephant.* International Elephant Foundation.

Poole, J. H. (1982). *Musth and male-male competition in the African elephant* (Doctoral dissertation). United Kingdom: University of Cambridge.

Poole, J. H. (1987). Rutting behavior in African elephants: The phenomenon of musth. *Behavior, 102,* 283–316.

Poole, J. H. (1989). Announcing intent: The aggressive state of musth in African elephants. *Animal Behavior, 37,* 140–152.

Poole, J. H. (1994). Sex differences in the behavior of African elephants. In R. V. Short & E. Balaban (Eds.), *The differences between the sexes* (pp. 331–346). United Kingdom: Cambridge University Press.

Poole, J. H. (1996.) The African elephant. In K. Kangwana (Ed.), *Studying elephants* (pp. 1–8). Nairobi, Kenya: African Wildlife Foundation.

Poole, J. H., Lee, P. C. Njiraini, N. & Moss, C. J. (2011). Competition and musth: A long-term perspective on male reproductive strategies. In C. J. Moss, C. Harvey, & P. C. Lee (Eds.), *The Amboseli elephants: A long-term perspective on a long-lived mammal* (pp. 224–237). IL: University of Chicago Press.

Roocroft, A., & Zoll, A. T. (1994). *Managing elephants: An introduction to their training and management*. Ramona, CA: Fever Tree Press.

Schulte, B. A. (2000). Social structure and helping behavior in captive elephants. *Zoo Biology, 19*, 447–459.

Short, R. V., Mann, T., & Hay, M. F. (1967). Male reproductive organs of the African elephant. *Journal of Reproductive Fertility, 13*, 517–536.

Slotow, R., Balfour, D., & Howison, O. (2001). Killing of black and white rhinoceroses by African elephants in Hluhluwe-Umfolozi Park, South Africa. *Pachyderm, 31*, 14–20.

Slotow, R., & Van Dyk, G. (2001). Role of delinquent young "orphan" male elephants in high mortality of white rhinoceros in Pilanesberg National Park, South Africa. *Koedoe, 44*, 85–94.

Slotow, R., Van Dyk, G., Poole, J., Page, B., & Klocke, A. (2000). Older bull elephants control young males. *Nature, 408*, 425–426.

Stokke, S., & du Toit, J. T. (2002). Sexual segregation in habitat use by elephants in Chobe National Park, Botswana. *African Journal of Ecology, 40*, 360–371.

Taylor, V. J., & Poole, T. B. (1998). Captive breeding and infant mortality in Asian elephants: A comparison between twenty Western zoos and three Eastern elephant centers. *Zoo Biology, 17*, 311–332.

Chapter 12

THE FIRST BACHELOR GROUP OF AFRICAN ELEPHANTS IN A NORTH AMERICAN ZOO

Lara C. Metrione, Patrick Flora, William Foster, Linda M. Penfold

A VISION FOR BACHELOR GROUPS OF AFRICAN ELEPHANTS IN ZOOS

Bachelor groups of African elephants (*Loxodonta africana*) occur naturally in the wild (Chiyo et al., 2011; Evans & Harris, 2008; O'Connell-Rodwell et al., 2011), and zoos and conservation facilities often strive to replicate natural social conditions when managing animals as a method to promote well-being overall. In addition to maintaining bulls in a natural social setting, housing African elephants in bachelor groups would maximize finite holding space, allowing group versus individual housing. Ideally, bachelor groups of bulls based at several zoos across the country could serve as regional repositories for bulls: young bulls could be sent to "bull school," where they would participate in a socially appropriate bachelor group, and mature bulls could move in and out of these facilities as required by breeding recommendations. This system of regional bull management would mirror wild African elephant bachelor groups in which bulls move in and out as they periodically enter musth and seek female groups for breeding opportunities (Poole, 1994; Poole & Moss, 1981). Such a system in zoos would have the further potential advantage of maintaining the bulls' "novelty" when they are introduced or reintroduced into a female group for breeding, possibly promoting cyclicity and pregnancy.

CREATING AND MAINTAINING A BACHELOR GROUP

Introductions

The first step in achieving this vision for bull management is to create successful bachelor groups at individual zoological facilities. The Birmingham Zoo in Alabama recently undertook management of four bull African elephants as a single group, a first among North American zoos. Gradually, three adult ("A," "B," "C") and one subadult ("D") African bull elephants were introduced. Bull A (age 31) arrived at Birmingham in December 2010 and was joined two months later by bull B (age 11). Bull C (age 11) arrived three months thereafter, and finally, almost a year after C's arrival, bull D (age six) arrived in April 2012.

Upon introduction to the program, bulls were rotated into different stalls (0.01–0.03 ha) each night to allow all of them to see and acclimate to one another. Each morning, staff interpreted traces of evidence to understand how the bulls fared the previous night. For example, the amount of hay the bulls left behind could indicate an unsettled night that resulted in more hay left uneaten. The location of manure piles, whether near the adjacent stall or in the center of the stall, could suggest time spent with the neighboring bull or as far as possible from neighbors, respectively. This information was then used to direct when and with which bulls introductions might begin. Initial introductions usually occurred in the stalls and then in a boma (0.02 ha) as a precaution in case intervention would be absolutely necessary. Based on outcomes of those brief introductions, bulls A and B or B and C could be exhibited together in the yard (1.82 ha) beginning in June 2011, and then all three were exhibited together for the first time in June 2012. By December 2012, when formal behavioral observations were initiated, A, B, and C were exhibited as a triad regularly and in all dyads. Meanwhile, bull D was exhibited first with A in June 2012. Next, during introductions between C and D, C exhibited behaviors including driving and chasing, mounting, tusking, and tense trunk over back (Table 1). To prevent these from escalating, D was first put on exhibit with C in the presence of A, the dominant bull, in February 2013. Similarly, bulls B and D were introduced in the presence of A shortly before all four bulls were exhibited together in May 2013.

Table 1

Ethogram for Behavioral Observations of Bull African Elephants at Birmingham Zoo

BEHAVIORAL CATEGORY	SPECIFIC BEHAVIOR	DEFINITION
Affiliative	Approach relaxed	Actor moves slowly toward recipient within two body lengths, with head low and ears lying against side of head
	Rubbing	Actor rubs head or body on recipient
	Sharing food/ object	Elephants share a pile of food or an object
	Sweeping	Actor "sweeps" two-thirds of trunk over back or head of recipient with relaxed motion
	Trunk tangle	Actor reaches out and loosely entangles trunk of recipient
	Trunk touch	Actor touches trunk to recipient
	Trunk toward	Actor extends trunk toward recipient

BEHAVIORAL CATEGORY	SPECIFIC BEHAVIOR	DEFINITION
Aggressive	Approach with head high, ears perpendicular	Actor marches toward recipient with head above shoulders and ears out perpendicular
	Charge	Rapid, forward lunging or rapid gait by actor with head high and ears perpendicular toward recipient
	Chase	Actor rapidly pursues recipient, who is moving away from the actor
	Drive	Actor places head or tusks on hindquarters of recipient, resulting in a displacement of recipient by actor of at least two body lengths
	Ears perpendicular	Actor's ears extended out perpendicular from the body, oriented toward recipient
	Ear-waving	Movement starts at the top of ear and moves to bottom, like a wave
	Head-butt	Actor butts head or trunk against recipient
	Headshake	Actor's head held above shoulders and shaken vigorously from side to side
	Kick	Actor kicks at recipient
	Mount	Actor raises forelegs and places them along either side of or on spine of recipient

BEHAVIORAL CATEGORY	SPECIFIC BEHAVIOR	DEFINITION
Aggressive	Push	Actor contacts recipient with enough force to displace recipient
	Push down on head	Actor pushes down on head of recipient with the base of the trunk, tusks, or mouth
	Spar	Elephants face each other head to head, pulling and pushing
	Tense trunk over back	Two-thirds or more of actor's trunk held tensely over back or head of recipient
	Throw	Actor throws object toward recipient
	Tossing trunk	Actor throws trunk up and/or out with force in direction of recipient
	Trunk slap	Actor sharply contacts recipient
	Turn toward with head high, ears perpendicular	Actor turns toward recipient with head above and ears perpendicular
	Tusking	Actor contacts recipient with tusk

BEHAVIORAL CATEGORY	SPECIFIC BEHAVIOR	DEFINITION
Subordinate	Back	Stepping back from opposing elephant
	Back into	Walking backward into another elephant so that rump contacts other elephant
	Flinch	Dodge or abruptly shift weight away from opposing elephant
	Look away	Quickly turning head away from opposing elephant
	Lower head or ears	Quickly drop head or ears in response to another elephant
	Move away	Yielding ground and moving away as a result of actions or encroachment by another elephant
	Protest rumble or roaring rumble	Low rumbling emitted when elephant is in conflict with another
	Stop	Elephant stops approaching opposing elephant as a result of an action by opposing elephant
	Turn away	Turn body away from approaching or threatening elephant
	Tie	Both elephants make a subordinate movement or neither makes a subordinate movement although both have reacted to the conflict (e.g., both hold their ground at the end of a spar)

MONITORING

Early in planning Birmingham's program, management recognized that behavioral and endocrine monitoring would be important in understanding how the bulls acclimated to the group. In wild bachelor groups, aggression is higher in the absence of a strict dominance hierarchy, especially among lower-ranking bulls (O'Connell-Rodwell et al., 2011), so it was important to evaluate dominance among the bulls. Dominance among wild bulls usually is determined by size (Poole, 1989), but during musth, testosterone concentrations (Ganswindt, Heistermann, & Hodges, 2005; Ganswindt, Rasmussen, Heistermann, & Hodges, 2005; Hall-Martin & Van der Walt, 1984) and aggression (Ganswindt, Heistermann, et al., 2005; Poole, 1989) increase, and musth bulls become behaviorally dominant over non-musth bulls regardless of body size (Poole, 1989). At Birmingham, it was unknown whether the constant all-bull environment might suppress musth and how a dominance hierarchy might develop in a zoo-housed bachelor group where space is finite, competition for resources is limited, and there are no females. Quantifiable behaviors (Table 1) described in previous ethograms (Burks et al., 2004; Freeman, Schulte, & Brown, 2010; Poole, 1987, 1989) were recorded approximately every other month for one year to measure dominance and subordination, frequency of aggressive and affiliative behavior, and frequency of initiated social interaction.. Dominance was determined by calculating the percentage of the total antagonistic interactions that resulted in a "win" (i.e., the interaction ended with the opponent behaving subordinately) for each bull in every possible dyad.

In addition, serum testosterone concentrations were monitored by enzyme immunoassay for evidence of musth and potential relationships with dominance. Reduced behavioral interactions and increased production of glucocorticoids can be characteristic of a "conservation-withdrawal" response to stressful stimuli (Carlstead, 1996); therefore, serum cortisol concentrations were monitored and their relation to the bulls' willingness to initiate social interactions was examined. The need to monitor these weekly serum hormone concentrations reinforced the importance of the training program in which all four bulls participated. Daily training sessions for behaviors used in medical procedures, bathing, and semen collection, which stretch and exercise muscle groups and which encourage patience and obedience, were an important tool for the staff to assess the day-to-day health and demeanor of each bull. These training sessions also provided the bulls with time when their minds were concentrated on structured tasks. Elephant cooperation during training was rated 1 through 5 by elephant keepers throughout the day, and ranged from unfocused, uncooperative behavior (1) to exceptionally responsive, cooperative behavior (5). Temporal gland secretion (TGS: Poole, 1987) also was rated daily: zero indicated no TGS and four indicated maximum TGS extending to the base of the lower jaw. Daily ratings enabled staff to note subtle changes in behavior patterns before more dramatic demonstrations of behavior (i.e., the transition to musth).

OUTCOMES TO DATE

Among wild bachelor groups, non-musth males engage in frequent aggressive interactions (~5–30/h: O'Connell-Rodwell et al., 2011), and sparring is important for the appropriate socialization of young bulls and dominance hierarchy enforcement (Evans & Harris, 2008). Thus, observations of sparring in this zoo-housed group came as no surprise, and the bulls differed in the degrees to which they interacted with each other and used aggressive and affiliative behavior. Bull C avoided interactions with A, the indisputably dominant bull, in particular. It was not until seven months into the monitoring process, about one month after the first time that all four bulls were exhibited together, that bull C really began to approach other bulls more often. Before all four bulls were exhibited together, the more interactive and more often aggressive bull B sought to interact heavily with C. Once bull D was brought into the mix, however, he and C occupied each other, which appeared to afford bull C the luxury of approaching the others in his own time and on his own terms. Eventually, by month 12, C even permitted and initiated affiliative contact with A.

Although bull C did appear to gain confidence, the group's dominance hierarchy remained consistent, and he was never more dominant than bull B or A, who was in the alpha position. The youngest and smallest bull, D, was the most subordinate and also had significantly lower testosterone ($p<0.05$) than did the others. Interestingly, testosterone concentrations did not differ ($p>0.05$) among the three older bulls, and higher serum testosterone concentrations corresponded with the alpha position in the dominance hierarchy only during bull A's musth. Similarly, there is no relation between rank and fecal testosterone metabolites among wild bull groups that have linear dominance hierarchies (O'Connell-Rodwell et al., 2011).

Although bull A did not engage in interactions often and usually exhibited affiliative behaviors, he maintained his position by winning the vast majority of the aggressive interactions in which he was involved. Lower frequencies of social behavior, including greeting in particular, were observed among wild bulls equal to or older than 36 years of age that played integral roles in bull social groups compared to younger animals (Evans & Harris, 2008), so bull A's generally passive behavior except when asserting his dominance would seem appropriate. The dominance hierarchy in this bachelor group, in which the oldest bull was alpha, could be important in preventing the natural, expected level of aggression from becoming excessive or destructive (Slotow, van Dyk, Poole, Page, & Klocke, 2000), as aggression among wild subordinate bulls increases when the structure of the dominance hierarchy degenerates (O'Connell-Rodwell et al., 2011). While size generally dictates dominance among non-musth bulls in the wild (Poole, 1989), bull B (2,823.6 kg), who engaged in more interactions and used aggressive behavior more frequently, was dominant over bull C (3,839.7 kg), who engaged in fewer interactions and used affiliative behavior more often. Thus, social activity and aggression seemed to play a role in determining dominance in this zoo-housed group.

The keepers' management strategies also certainly could have influenced social interactions among the bulls. For example, bulls B, C, and D were given access to the main yard first, with time to disperse away from the gate and explore while bull A received his bath. Once A entered the yard, the other bulls had the option of remaining apart from A or moving to interact with him. Even in 2011, prior to the beginning of formal behavioral monitoring, because C appeared nervous upon first introduction to A in the barn, managers allowed him to form a relationship with B first, then introduced A to C in the presence of B. Throughout the process of group formation, depending on keepers' perceptions of bull demeanor during the day, stall assignments for the night were based on fostering confidence-building interactions between the bulls. For example, if bull C used aggressive behavior frequently with D during the day, D might be housed between A and C so that A could mediate interactions and provide a refuge on one side of the stall.

Importantly, none of the bulls showed sustained, elevated cortisol, which can be indicative of stress, and cortisol concentrations in these bachelors were consistent with those measured in a bull at another facility housed in the vicinity of females but no other males. Throughout the study, cortisol concentrations were significantly lower for bulls B and D ($p<0.05$), whose concentrations were similar ($p>0.05$), compared to bulls C and A, whose concentrations also were similar ($p>0.05$). Neither dominance nor age accounted readily for these similarities and differences. Both of the bulls that initiated social interactions more often had lower cortisol than did the bulls that initiated social interactions less often. Although this outcome could be coincidental, it is conceivable that bulls who tend to seek social stimulation might be less sensitive to social stimuli that could trigger an adrenal response than bulls who behave more passively.

CONCLUDING REMARKS

The bachelor group of African elephants at the Birmingham Zoo has been remarkably successful. This is attributable in large part to the careful introduction process, rigorous training program, excellent facilities, and dedicated staff. Behavioral and endocrine monitoring helped define social dynamics in the group, indicate changes in behavior patterns, and provide a measure of how the bulls responded physiologically to their environment. Wild bachelor groups are the model for the social environment that we would like to facilitate in our zoos, and provide a basis for our expectations of zoo-housed bachelor elephants. However, a methodical monitoring plan allows managers to be open and responsive to important differences from one bachelor group to the next, thereby ensuring that the well-being of those bulls is at the heart of the management system.

ACKNOWLEDGMENTS

We would like to thank the elephant care team and health center staff at the Birmingham Zoo, as well as the Jacksonville Zoo and Gardens elephant care team and vet hospital staff. We also thank Cayman Adams for performing the endocrine assays.

REFERENCES

Burks, K. D., Mellen, J. D., Miller, G. W., Lehnhardt, J., Weiss, A., Figueredo, A. J., & Maple, T. L. (2004). Comparison of two introduction methods for African elephants (*Loxodonta africana*). *Zoo Biology, 23*, 109–126.

Carlstead, K. (1996). Effects of captivity on the behavior of wild mammals. In D. G. Kleiman, M. E. Allen, K. V. Thompson, S. Lumpkin, & H. Harris (Eds.), *Wild mammals in captivity: Principles and techniques* (pp. 317–333). IL: University of Chicago Press.

Chiyo, P. I., Archie, E. A., Hollister-Smith, J. A., Lee, P. C., Poole, J. H., Moss, C. J., & Alberts, S. C. (2011). Association patterns of African elephants in all-male groups: The role of age and genetic relatedness. *Animal Behavior, 81*, 1,093–1,099.

Evans, K. E., & Harris, S. (2008). Adolescence in male African elephants, *Loxodonta africana*, and the importance of sociality. *Animal Behavior, 76*, 779–787.

Freeman, E. W., Schulte, B. A., and Brown, J. L. (2010). Using behavioral observations and keeper questionnaires to assess social relationships among captive female African elephants. *Zoo Biology, 29*, 140–153.

Ganswindt, A., Heistermann, M., & Hodges, K. (2005). Physical, physiological, and behavioral correlates of musth in captive African elephants (*Loxodonta africana*). *Physiological and Biochemical Zoology, 78*, 505–514.

Ganswindt, A., Rasmussen, H. B., Heistermann, M., & Hodges, J. K. (2005). The sexually active states of free-ranging male African elephants (*Loxodonta africana*): Defining musth and non-musth using endocrinology, physical signals, and behavior. *Hormones and Behavior, 47*, 83–91.

Hall-Martin, A. J., & Van der Walt, L. A. (1984). Plasma testosterone levels in relation to musth in the male African elephant. *Koedoe, 27*, 147–149.

O'Connell-Rodwell, C. E., Wood, J. D., Kinzley, C., Rodwell, T. C., Alarcon, C., Wasser, S. K., & Sapolsky, R. (2011). Male African elephants (*Loxodonta africana*) queue when the stakes are high. *Ethology, Ecology & Evolution, s*, 388–397.

Poole, J. H. (1987). Rutting behavior in African elephants: The phenomenon of musth. *Behavior, 102*, 283–316.

Poole, J. H. (1989). Announcing intent: The aggressive state of musth in African elephants. *Animal Behavior, 37*, 140–152.

Poole, J. H. (1994). Sex differences in the behavior of African elephants. In R. V. Short & E. Balaban (Eds.), *The differences between the sexes* (pp. 331–346). United Kingdom: Cambridge University Press.

Poole, J. H., & Moss, C. J. (1981). Musth in the African elephant (*Loxodonta africana*). *Nature, 292*, 830–831.

Slotow, R., Van Dyk, G., Poole, J., Page, B., & Klocke, A. (2000). Older bull elephants control young males. *Nature, 408*, 425–426.

Chapter 13

HERD LIFE

MEETING THE PHYSICAL AND BEHAVIORAL NEEDS OF ASIAN ELEPHANTS THROUGH EXHIBIT DESIGN

Elizabeth S. Herrelko and Tony Barthel

INTRODUCTION

The physical and psychological needs of elephants have been at the forefront of modern zoo management efforts for many years (e.g., Sukumar, 1992; Fowler & Mikota, 2006; Kurt & Garai, 2006; Wemmer & Christen, 2008; Greco et al., 2016; Meehan, Mench, Carlstad & Hogan, 2016). Complex habitats are required to meet the needs of these intelligent, large, social species. In recent years, great strides have been made in group structure and habitat design around the globe (e.g., Dublin Zoo, Saint Louis Zoo, Denver Zoo, Houston Zoo, Dallas Zoo, Oregon Zoo, Birmingham Zoo, Disney's Animal Kingdom, Busch Gardens, Kölner Zoo, Zoo Zurich, and Zoo Copenhagen). With the addition of each new exhibit, the zoological field continues to readily share ideas to improve standards and raise the status quo together.

Megafauna are a large focus within the zoological community (Hancocks, 1995), and elephants in particular have been the subject of management and policy changes (e.g., the Association of Zoos and Aquariums' accreditation standards regarding collection sizes: AZA, 2016) designed to eliminate singly housed elephants and facilitate zoos' ability to achieve breeding success to stabilize the population. Some might say these changes were the result of social pressures, but others might say those pressures advanced what elephant people had wanted to do for years. In reality, it was likely a combination of the two.

A NEW REFLECTION OF THE ZOO'S COMMITMENT TO ELEPHANTS

The Smithsonian's National Zoological Park (NZP) has undergone a significant transformation in recent years to transition the pachyderm and hoof stock collection plan into a dedicated resource for elephants. After a 10-year commitment to the design and construction of a new habitat driven by the physical and behavioral needs of our elephants, NZP opened Elephant Trails in 2013, a facility based on elephant needs and natural history, and intended to encourage positive well-being.

For many years (1930s–2000s), the NZP elephant house was home to pygmy hippos, giraffes, greater one-horned rhinos, Asian and African elephants, and Nile hippos. Much like many facilities at the time, all of these species lived in part of what is now considered Elephant Trails. The house was built in the early 1930s and, with a few exceptions, had only changed cosmetically by the year 2000. Most of the space was dedicated to visitors. By 2013, the visitor-to-animal space ratio changed, and elephants were provided more square footage (Figure 1).

FIGURE 1: PHOTOGRAPHS OF FORMER PACHYDERM HOUSE TAKEN IN (A) 1930s, (B) 1990s, AND (C) 2016 SHOWING THE PROPORTION OF VISITOR SPACE TO ANIMAL SPACE OVER TIME. THE VANTAGE POINT OF EACH PHOTOGRAPH IS FROM THE CENTER OF THE BUILDING, AND THE DOMED AREA IS IN THE SAME LOCATION IN EACH PHOTOGRAPH.

The transition was not just a matter of space; in fact, space was not a significant driving factor with respect to size overall. Research with other species has suggested that size itself is not the most important property of space (de Waal, 1989; Ross, Calcutt, Schapiro, & Hau, 2011; Herrelko, Buchanan-Smith, & Vick, 2015; Meehan et al., 2016). For elephants, space needs are dictated by other key factors, such as potential numbers of elephants, need for variety, having elements of choice and control, and ways in which they might use the space not only as individuals, but as a group. This focus evolved from managing the individual to managing a group structure, a herd. NZP wanted to move from an old, outdated, and insufficient living environment to one driven by the animals' needs and natural history.

In addition to establishing the need to improve the facility for the elephant residents, NZP determined that the facility also had to be improved to help deliver a consistent message to its visitors. For example, the indoor elephant facility at NZP consisted of a series of connected stalls visible to the public (Figure 2). Typically, zoo guests experienced this as a series of side-by-side boxes and could not intuitively comprehend how that met the elephants' needs. The image painted a certain picture in their minds. Even if the elephants had a choice to move from one place to another and the exhibit could meet certain needs (e.g., Olson, unknown), it did not matter, as all the visitors could see was an elephant in a small room, often perceived as an "elephant in a box."

FIGURE 2: INDOOR ELEPHANT ENCLOSURES (A, B) IN THE PREVIOUS PACHYDERM HOUSE.

Zookeepers spent so much time rationalizing the appearance of the exhibit that they often were unable to say what they wanted to about the animals. The opportunity to communicate the crucial link between the zoological experience and animal behavior and conservation was challenging. It was difficult to teach zoo guests about elephants and inspire action when those conversations often had to begin with a lengthy explanation of the way in which we thought the existing facility provided for the elephants' needs. As NZP's program focused more and more on the elephant herd and the importance of a rich social life for all elephants, the image of a series of box-like rooms designed clearly for one elephant each did not help tell that story.

Ultimately, NZP created an exhibit in which elephants had access to interconnected habitats and a variety of substrates; it was a place in which they had to make choices not only about how to use the exhibit, but also how, where, and when to socialize. During the design phase, the zoo had several goals with a common element: create an exhibit based on elephants' physiological and behavioral needs to elevate basic husbandry and create our own best practices.

HABITAT GOALS

The elephant team focused on: (1) sociality when considering spaces, access, and exhibit flow; (2) space, flexibility, and innovative concepts to facilitate exercise; (3) substrate to promote good foot and joint health; (4) enrichment to stimulate cognitive processes; and (5) opportunities for choice to allow elephants to derive as much benefit from the facility as possible, dependent upon individual and herd needs. These goals can be discussed as separate concepts, but as the following sections highlight, they are linked inextricably.

SOCIALITY

It is clear through our interactions with elephants, the topical literature (e.g., Moss & Poole, 1983; Lee, 1987; Archie, Moss, & Alberts, 2006; Evans & Harris, 2008; de Silva & Wittemyer, 2011; Fishlock & Lee, 2013; Prado-Oviedo et al., 2016), and the subjects discussed at the Elephant Wellness Workshop (2016) that sociality is the critical element in elephant life. As with every animal, their history shaped who they are today. Understanding the culture and story of the NZP elephants (using anthropomorphism to help the reader relate to the way the keepers experienced it) will help explain the role of sociality within this exhibit.

ORIGINAL HERD

Before the initial discussions about a new exhibit, the herd consisted of African elephant Nancy and Asian elephants Shanthi, Ambika, and Toni. Nancy was the matriarch and leader of the group, and although she passed away in 2000, her influence is still seen within the herd. Her death affected the current group dynamics, and, as described among wild elephants (Douglas-Hamilton, 1998; Moss, 1988), they had to adjust after she died. Without a clear matriarch (more typical of African elephants), the remaining Asian elephants exhibited a nonlinear social hierarchy. Although this is typical for the species (de Silva, Volker, & Wittemyer, 2016), it was a substantial change for this group. Shortly after Nancy died, Kandula (son of Shanthi) was born in November 2001. At this point, a rich social life developed around him. When Toni died in 2006, social dynamics adjusted again, and that left Shanthi mediating between a young, boisterous male and a geriatric female (Figure 3).

FIGURE 3: ASIAN ELEPHANTS AT THE SMITHSONIAN'S NATIONAL ZOO, CIRCA 2011. THE BEHAVIOR SEEN IN THESE PICTURES SUMS UP THE SOCIAL RELATIONSHIP AT THE TIME: (A AND B) KANDULA (LEFT, A YOUNG BULL) IS SEEN DIRECTING AN ANTAGONISTIC INTERACTION TOWARD AMBIKA (RIGHT, FEMALE IN HER 60S) AROUND HIS MOTHER, SHANTHI (MIDDLE, FEMALE IN HER LATE 30S) BEFORE (C) AMBIKA PUT HIM IN HIS PLACE.

The zoo had grand plans of creating a home for a large herd and building it slowly over time, but cows were not becoming pregnant, and finding additional reproductively viable females was not going well. The reality of the North American elephant population did not match that herd vision. With no calves, the elephant team began to search for additional adult females to join the herd to ensure companionship and help build a richer social experience for the elephants at NZP. Consequently, the elephant team recognized that incorporating animals and/or groups with varied life histories into the NZP collection would require considerable flexibility in management and facility use to accommodate the needs of all the animals as they established new relationships. NZP was able to accommodate this type of change because the exhibit was designed with sociality in mind (in terms of space, access, and exhibit flow and flexibility).

In 2013, shortly after opening the exhibit, the keepers welcomed Bozie (Figure 4). Bozie came to NZP from Baton Rouge Zoo in Louisiana after the loss of her herd mate. Following a brief quarantine period upon arrival, her introduction to the resident herd was accelerated because of everyone's behavior. Bozie clearly was interested in being near the other elephants, and they, in turn, showed interest in her. The management process went very quickly from howdy introductions, in which individuals have visual, auditory, olfactory contact, and physical contact through a barrier, to full introductions with physical contact and no barrier present. There was a small amount of pushing and shoving as the new herd members tested one another. Ambika, despite her old age, met her in the doorway and then backed down as if to indicate that she was not a pushover but also not a threat. Immediately after the first full introduction, Bozie resided full-time with Ambika and Shanthi. Ambika is an animal who is slow to trust, and although tolerant of Bozie's new presence, she did not embrace it initially. She stopped lying down at night, as she had done in the past when she did not seem to trust another elephant in her area. Relationships improved over the first year, and then everything changed when the Calgary herd arrived.

FIGURE 4: ASIAN ELEPHANT BOZIE (FEMALE IN HER 40S).

CALGARY HERD

In 2014, NZP welcomed three female elephants from Calgary Zoo: Maharani (also known as Rani, in her 20s); Kamala, Rani's mother (in her late 30s); and Swarna (in her late 30s: Figure 5). This herd presented its own challenges. Kamala and Swarna were bonded closely, but at some point after Rani was born, the mother/daughter duo began to team up against Swarna, often displacing her, pushing her, and even knocking her around. Animal management was adjusted to mitigate those challenges, but fights occurred daily, and Swarna often would go down to her knees to avoid being knocked down.

Initial assumptions were that Rani instigated the aggression largely and that Kamala became involved later. However, as the team came to know them better, it became clear that, at least at NZP, Kamala was more often the instigator and Rani the follower.

FIGURE 5: ASIAN ELEPHANTS (A) MAHARANI, AKA RANI; (B) SWARNA; AND (C) KAMALA.

OFF TO OKLAHOMA

Time went on, and in 2015, Kandula moved to Oklahoma City Zoo; his move was based on a Species Survival Plan (SSP) recommendation and served as an excellent opportunity to transfer him to a wonderful facility and program. The move gave him many new social opportunities and allowed him to be represented well within the genetic population. The process to move him to another facility was a smooth one: As Kandula matured, he became increasingly independent from Shanthi and Ambika, his natal herd. It was a gradual process driven by the elephants and set them up nicely for transport. The elephant team planned social grouping decisions based on elephant behavior to mimic the way animals would experience it in the wild; the team did not want to do it any other way.

CURRENT HERD

Currently, NZP manages six elephants as two subgroups within a single herd. All six are involved completely in one another's lives; they are aware of and vocalize to one another; they are not all physically together yet, but that will come in time. Now, Swarna (originally part of the Calgary group) participates successfully in both subgroups. She spends her days with Ambika, Shanthi, and Bozie, and evenings with Kamala and Rani. This gives her the opportunity to not be bullied constantly and the time to build social skills in other ways. One of the amazing outcomes of this social change was that it changed her dynamic with Kamala and Rani, and they no longer bullied her.

Based on what zoo staff knew about each elephant's temperament and the way they acted during the initial introduction period, keepers expected the encounters to go smoothly when Swarna was introduced to Shanthi and Ambika without any physical barriers between them. Unfortunately, the reality was just the opposite, and Swarna became aggressive to Ambika and Shanthi quickly. She began to push them, knock into them, and even knocked Ambika, then in her late 60s, to the ground. This behavioral display was a surprise because historically, she had been submissive and bullied by other elephants. As such, she was expected to be guarded and reticent rather than confident and antagonistic.

Could this be an example of behavioral cusps (Rosales-Ruiz & Baer, 1997) in action? Swarna began to spend more time with Ambika, Shanthi, and Bozie. After a while, when she spent the night with Rani and Kamala, she was no longer the recipient of aggression. It was a different situation, a different relationship within their subgroup. As Morris and colleagues discussed (Morris et al., this volume), she may have been learning additional skills (e.g., behavioral methods or rules) when interacting with other elephants that she was able to apply to the antagonistic herd mates; if so, her interactions with Ambika, Shanthi, and Bozie's subgroup had far-reaching consequences. Alternatively, perhaps Rani and Kamala's behavior changed because Swarna was absent for a portion of the day. It is difficult to tease out the reason for the outcome, but based on their observations, care staff suspect it was more of the latter.

Animals are amazing, and challenges bring out incredible behaviors in other individuals. Bozie began to exhibit classic matriarchal behavior. She would not tolerate Swarna's hostile behavior. She positioned herself between Swarna and the other elephants, controlled every one of Swarna's movements, and mediated when Swarna appeared to be preparing for an antagonistic interaction with either Shanthi or Ambika. Because of her interventions, all of the elephants began to lie down and sleep together. This seemed to be the final key in building trust and also served as a good indicator of their behavioral status.

Group structure is the crux of what is important for an exhibit. It was particularly helpful that our exhibit allowed the current herd to have opportunities for fission-fusion. It also was important to consider hierarchies and the pressure they exert on the lower-ranking members of the group. Enclosure design set this group (elephants and keepers) up for success. For example, with multiple doors to each enclosure, one dominant animal could not block all access points, a situation that would put unnecessary stress on the social group. The lessons learned during the development of the current herd helped transition the way the animal care team saw the future of elephant management at NZP.

SPACE

Space does not need to be the central issue in exhibit design, but it is an important consideration. When considered carefully, space can have a profound effect on sociality. Therefore, building interconnected habitats provides space to choose. From a cognitive perspective, the elephants have to think about where they want to go, who they want to be with (or not), and what they choose to do.

OUTDOORS

The outdoor exhibit consists of four habitats/yards with multiple access points and approximately 2 acres of usable space (Figures 6 & 7, Table 1). When we talk about "usable space," we are referring to "used space," in that the management of the herd (the "software" of keeper training, slow feeders, etc.) facilitates movement throughout the entire exhibit. There are three pools in the yards (Figure 8) with open bank lines that do not put extra social pressure on the group. From a design perspective, there are a number of reasons, such as maximizing space, pool depth, and aesthetic creativity, to create long linear pools that often have narrow entrance paths. However, this layout is less than ideal from an elephant's perspective. Bathing is often a social experience, or at least has the potential to become one. Narrow entrances that can be blocked easily can make an elephant feel insecure and less likely to enter the pool. By creating large open banks with gradual slopes by which the elephants can enter the pool, they tend to be more comfortable in the pool and more likely to choose to take a dip. Recirculating water was incorporated into one of the pools, thus allowing us to be better stewards of our environment (minimal water waste). A recirculating pool also is less labor intensive to maintain.

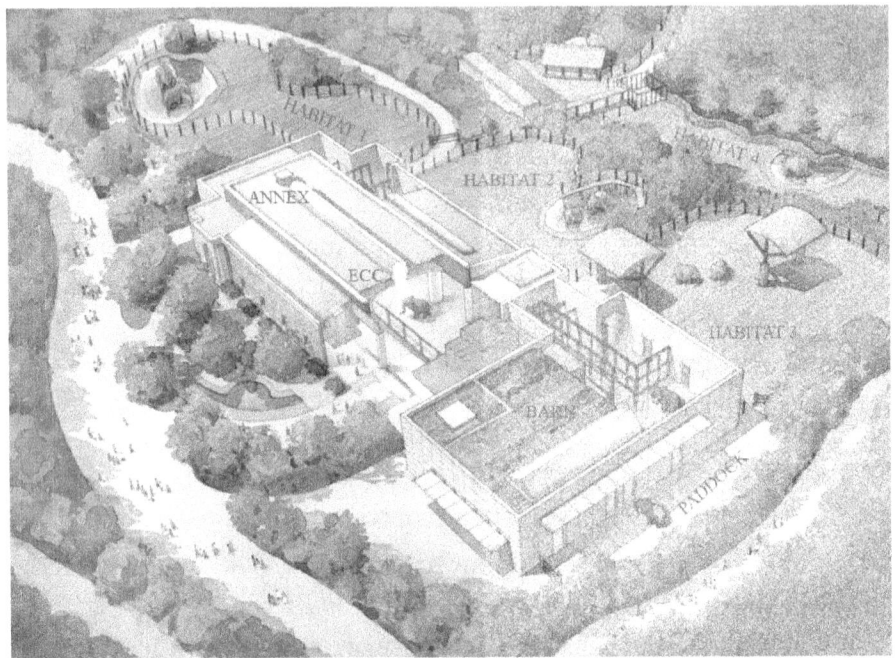

FIGURE 6: A SKETCH OF THE ELEPHANT TRAILS EXHIBIT
LAYOUT. THE TREK IS ACCESSED THROUGH HABITAT 4.

FIGURE 7: LAYOUT OF ELEPHANT TRAILS AS SHOWN BY GOOGLE MAPS.

Table 1

Enclosure Metrics

		ENCLOSURE SIZE (SQUARE FEET) SPACES	ENCLOSURE SIZE (SQUARE FEET) POOLS
Outside	Habitat 1	12,614	2,656
	Habitat 2	8,950	980
	Habitat 3	8,900	
	Habitat 4	35,098	2,521
	Paddock	3,734	
Inside	Stall 1	910	
	Stall 2	700	
	Stall 3	694	
	Stall 4	1,674	
	Stall 5	926	
	Stall 6	928	
	Stall 7	946	
	ECC*	9,245	Splash pool

*Elephant Community Center (ECC) includes square footage of the splash pool.

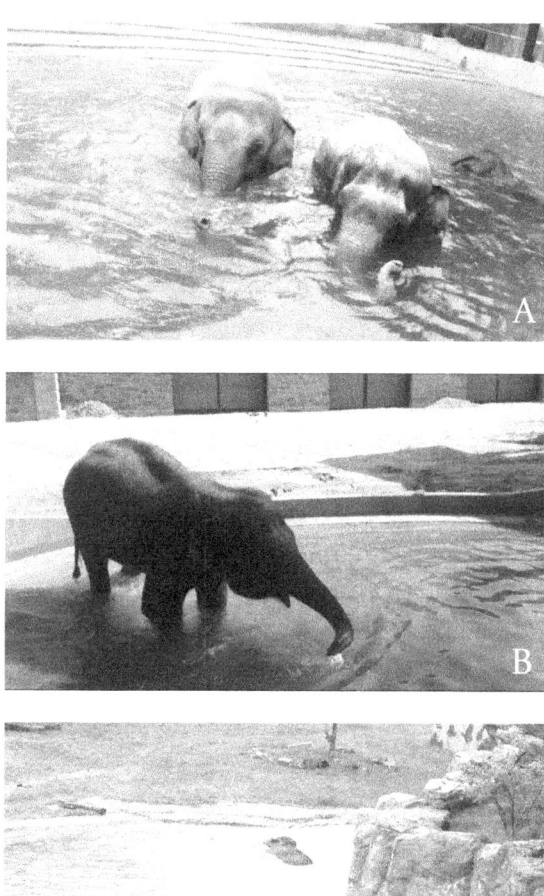

FIGURE 8: OUTDOOR ELEPHANT POOLS: (A) HABITAT
1, (B) HABITAT 2, AND (C) HABITAT 4.

The Trek, a quarter-mile pathway through the zoo with an elevation change of 51 feet, provides an excellent opportunity for exercise (Figure 9). It also offers a change of scenery for the elephants—when a nearby exhibit (American Trail) was being built, the elephants became extremely curious and watched the construction (Figure 10). When the neighboring exhibit was finished, unsuspecting visitors could catch a glimpse of elephants walking through the woods.

The original goal of the project was to have the trek loop back to another part of the outdoor enclosure, but unfortunately, it was value engineered out of the project. It is wonderful to see design ideas with treks that go somewhere, and we stress the importance of having a destination at the end of an enclosure feature such as this.

FIGURE 9: AMBIKA WALKING THROUGH THE TREK.

FIGURE 10: KANDULA PAUSING ON THE TREK WITH A VIEW OF NEARBY CONSTRUCTION.

INDOORS

The indoor exhibit consists of seven stalls, two transfer hallways, two elephant restraint devices (ERDs), and one community center, referred to as a herd room in other facilities (Table 1). The building was designed so each stall would have direct access to the adjacent stalls, the elephant transfer chute, and an outdoor area (Figure 11). This permits the elephants and staff extensive options in facility use. It also provides an effective method of reducing the social pressures related to dead ends.

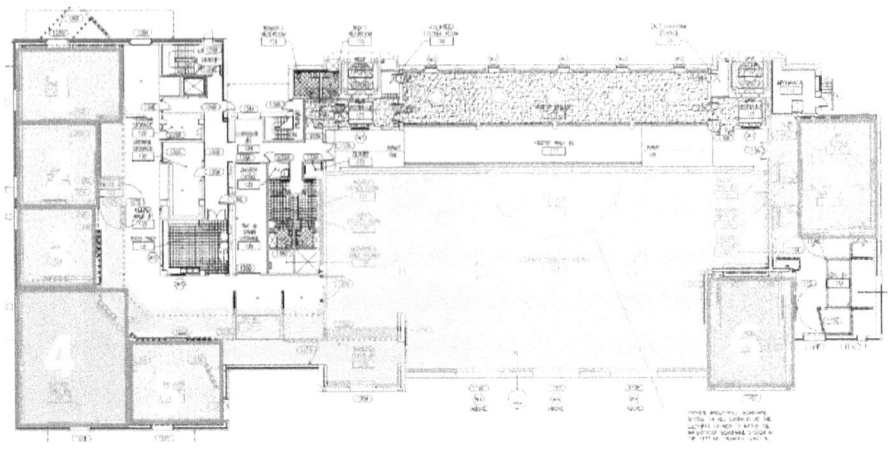

FIGURE 11: LAYOUT OF THE ELEPHANT TRAILS INTERIOR ENCLOSURE SPACE.
ELEPHANT STALLS ARE HIGHLIGHTED IN GREEN, TRANSFER HALLWAYS
IN BLUE, AND THE ELEPHANT COMMUNITY CENTER IN YELLOW.

The Elephant Community Center (ECC) is our largest indoor space and provides opportunities for large gatherings (Table 1, Figure 12). This space is the ultimate expression of the zoo's focus on designing for the elephant herd versus the individual elephants. Although it is a nice gathering space at any time of the year, it is particularly important during periods of inclement weather when the outdoors is not very inviting. Multiple elephants can share this space simultaneously while still having the freedom to move off on their own. The ECC features three large "bay" doors in the back that open on to an outdoor exhibit (Figure 13). When opened, the indoor space is simply an extension of that outdoor exhibit. The space is covered with 5 feet of sand, which provides both a natural substrate for the elephants and allows the elephant staff to change the topography of the space when desired. Also included is a small splash pool that is approximately 18 inches deep and can be filled with tempered water; there also are three showerheads above the pool that the elephants or the keepers can activate.

FIGURE 12: ELEPHANT COMMUNITY CENTER.

FIGURE 13: ELEPHANT COMMUNITY CENTER VIEW SHOWING THE
THREE "BAY" DOORS OPENING TO AN OUTDOOR EXHIBIT.

FEATURES

Elephant Trails is an LEED Gold Certified facility. Together with the use of recycled material during construction and highly efficient mechanical systems, the facility has other features that both make the building work better for the elephants and contributed to the LEED certification. Those include geothermal heating, operable skylights, and green roofs (Figure 14). Liquid-filled tubes run through the walls and floors of the facility. The liquid is heated in the winter and heats the floors and walls in turn. The heat is supplemented by a geothermal well system that heats the liquid in the tubes. In addition to its environmental benefits, this system has been found to keep the building from cooling too quickly in the winter when elephants are given access outside. By heating the mass of the building rather than just the air in the building, the space stays warmer longer, allowing keepers to give the elephants access outside even during cooler weather. The skylights over the enclosures are operable and programmed to open in the warmer months. This allows the hot air to escape through the open vents and draws cooler air into the building through the open doors, which circulate the air. The skylights also are equipped with automated shade cloths that close on hot, sunny days to shade the space below. Finally, the facility has green roofs in several areas. These rooftop gardens provide habitat for local wildlife, reduce water runoff, and, most importantly, provide excellent insulation. These systems work together to help keep the facility comfortable for elephants year-round.

FIGURE 14: SKYLIGHT ON THE ROOF OF THE ELEPHANT
COMMUNITY CENTER SURROUNDED BY A GREEN ROOF.

SUBSTRATE

By design, the majority of the substrate throughout the Elephant Trails facility is natural. This includes sand, grass, and soil. It also was deemed necessary to have some cleanable floors for veterinary purposes, so some of the enclosures have a rubber coating over the concrete slab. Prior to creating the Elephant Trails exhibit, NZP experimented for some time with surfaces other than concrete. Several different rubber floor options were tested, and ultimately, one of those was chosen for the new facility. Six years later, the rubber floor is still in good condition and has withstood the daily use and cleaning that comes with caring for elephants.

Before the construction of the new facility, we wanted to incorporate natural substrate floors indoors, but good models were difficult to find. Therefore, we began to experiment with sand floors. The only workable solution for the older facility was to build up the edges of the stalls to make a box so sand could be added on top of the concrete slab. We saw immediate benefits of having indoor natural substrates. In addition to the benefits we anticipated for the elephants' feet and joints, the sand also provided opportunities for the elephants to dust bathe inside, a natural behavior that had previously been limited only to the outside. However, this "sandbox" approach was also problematic. To minimize the need for the elephants to step up and down too much, the height of the box was 12 inches, which limited the depth of the sand within to approximately the same height. We found that the sand became soiled, wet, and compacted quickly, and had to be changed completely every six to eight weeks. Despite these problems, we were still committed to including natural indoor substrates in the new facility. Modeling on our experiences with sand in our outdoor paddocks, which did not need to be changed, we opted to build sand beds approximately 5 feet deep indoors with efficient drainage beneath. This approach has worked well; although it needs to be augmented periodically, we do not need to change the sand regularly. We have found that the sand increases the dust indoors significantly, but have been able to mitigate this largely with daily misting.

The outdoor yards are either sand, which like the indoor area needs to be augmented periodically, or grass grown in a sand bed with organics mixed in the top layer. We chose Bermuda grass because of its durability and maintain it with regular irrigation. The results have been very successful, as we have been able to keep grass in much of our new yards. In addition to providing natural grazing opportunities for the elephants, it creates a more attractive exhibit.

We have a mixture of substrates throughout the facility. The vast majority of the area, indoors and out, is covered with natural substrates, including sand, soil, and grass. Indoors, we also have the rubber floors and some concrete areas. In addition to the sand and grass outside, we have concrete pools and the trek, which is a paved surface. We believe this combination of substrates promotes both healthy nail and pad wear and good joint health.

ENRICHMENT

The primary goal of NZP's enrichment program is to enhance well-being by providing stimulating environments and management, giving opportunities based on natural abilities and behaviors, and allowing animals to exercise choice and control in their environment. The concept extends beyond putting toys in a yard and identifies what behaviors elephants need to exhibit to be elephants and then finds ways to allow them to engage in those behaviors. The Elephant Trails enrichment program can be broken down into several key categories including structural, elephant proof, sensory, food related, routine and change, socialization, training, and free choice enrichment. As we fine-tune the craft of elephant management, we consider these topics increasingly to be part of an animal's basic needs rather than additional components of husbandry (e.g., Burghardt, 1996). Whichever way you choose to look at it, these are concepts that help us reach our goal of supporting the elephants' physical and behavioral needs.

STRUCTURAL

Structural enrichment refers to the exhibit's ability to support objects that elephants can pull and manipulate with the weight of their bodies and strength without compromising the building's integrity. By incorporating this concept into the new exhibit design (not present in the previous building), elephant management opportunities increased. One example is the hoist beam trolley (Figure 15). All of the elephant stalls have a hoist beam equipped with a 12-ton hoist for emergency elephant care. These beams also are equipped with a trolley that can roll freely along the beams. Keepers hang large enrichment items from these trolleys to give the elephants opportunities to push them around. The behavioral goal is to encourage physical activity in addition to cognitive opportunities.

FIGURE 15: THE (A) HOIST BEAM SERVES AS STRUCTURAL ENRICHMENT THAT ALLOWS (B) KEEPERS TO SECURE LARGE, HEAVY OBJECTS ABOVE THE ELEPHANTS' HEADS AND ENABLES (C) ELEPHANTS LIKE KANDULA TO PLAY WITH ITEMS HANGING FROM THE TROLLEY THAT MOVE THE LENGTH OF THE STALL.

ELEPHANT-PROOF ADDITIONS

Elephant-proof additions are driven by the behaviors we want to promote by opportunities to manipulate large, heavy objects. Rather than limiting elephants with hot wire so they do not damage anything, we gave them more options by elephant-proofing everything within reach. NZP took this approach by intentionally including large, fixed scratching structures in the facility that the elephants can rub and scratch on, and making it possible to place large, movable objects throughout the facility for the elephants to manipulate (Figure 16). With this approach, staff can give the elephants free rein to push logs and tires without worrying that they might damage something.

FIGURE 16: AMBIKA MOVING A 3,000-POUND LOG STRUCTURE
BURIED DEEP WITHIN THE SAND SUBSTRATE.

SENSORY

Evans (this volume) discussed the importance of elephants' weird appearance: large ears and a long nose. The ways they hear, smell, and see are key factors in the way they live, and providing new ways for them to explore using those senses celebrates their morphology (Figure 17). Of course, just because you provide it does not mean they want it. When music is on, do they have the choice to turn it off? We know from past research that chimpanzees prefer Indian and African music to silence (Mingle et al, 2014) and that domestic cats prefer music that includes birds chirping or sounds similar to purring (Snowdon, Teie, & Savage, 2015), some primates prefer silence over music (McDermott & Hauser, 2007; Ritvo & MacDonald, 2016), and a study on singly housed baboons showed no effect on physiological or behavioral measures (Brent & Weaver, 1996). Considering choice as a factor in sensory enrichment can be just as important as the enrichment itself. Is it important for animals to be able to control these elements? Alternatively, if they are in a herd, what influence does a single animal's choice have on the entire group?

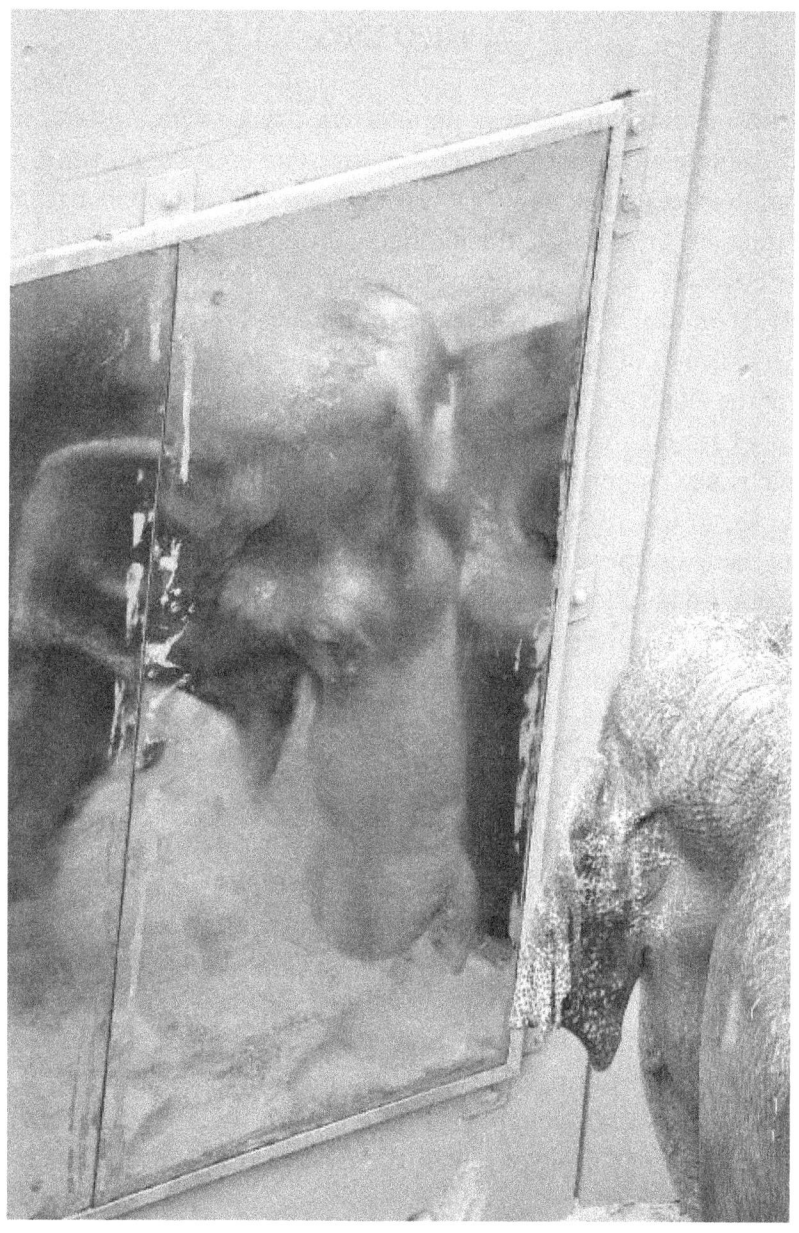

FIGURE 17: AMBIKA HANGING OUT WITH HER FAVORITE
ELEPHANT (MIRROR ENRICHMENT).

FEEDING

Feeding enrichment is always popular with those who are food motivated. We rely on it to prolong feeding time and create opportunities for animals to think about how to access it. One significant challenge facing elephant care programs is the method commonly used to feed the bulk of the elephants' diets. At NZP, grains and produce are target-fed to individuals more easily, but hay and browse are fed socially and offered in bulk to a group of elephants because foraging together is an important part of the elephants' social life. One of the problems with the system of offering hay and browse is that food is provided in a manner that allows individuals to monopolize large portions easily. This, in turn, adds stress to the social unit, does not promote exercise or movement, and minimizes the time spent foraging each day, which potentially can lead to boredom. An ideal feeding strategy would require the elephants to spend longer periods eating the same amount of food, move around more while foraging, and minimize food competition by providing plenty of feeding locations. This also would enhance the guest experience by creating interesting opportunities to watch elephants forage and interact; a key component that helps zoos create a powerful, memorable experience in an environment in which animals thrive enables them to educate people better and serve on the front lines of conservation (e.g., M. Sanjayan, personal communication, October 19, 2016).

At NZP, our goal is to have multiple hay and browse feeders of various types installed throughout the facility. The design of both would be focused on increasing foraging time, encouraging movement, spreading out feeding opportunities to decrease competition, and finally, making more efficient use of elephant keeper staff time. If the exhibit design can work well to meet elephant needs, the facility and the elephants themselves can serve as the bulk of their own enrichment.

ROUTINE AND CHANGE

Both routine and change affect animal behavior and welfare (Bassett & Buchanan-Smith, 2007). Care staff can change the time and delivery of food frequently to avoid negative anticipatory behaviors (e.g., Wilson, Bloomsmith, & Maple, 2004) and, at the same time, make special efforts to provide a sense of routine. For example, keepers might use different knocks before entering a room to help animals identify which keeper is about to come in (Rimpley & Buchanan-Smith, 2013). With respect to feeding, food might be delivered via a puzzle feeder, handfed, or hidden in an enclosure. These changes would keep animals guessing, while an element of routine remains the same, perhaps the signal that food is available. With respect to locomotion and shifting animals from one part of an exhibit to another, the layout of the exhibit could include path changes as requested by keepers. The concept of going from the outside to inside is the same, but the direction and movement differ. The routine comes in the form of the cues keepers use to direct behavior. These types of changes not only provide opportunities to vary the elephants' daily routine but also can prove invaluable when flexibility is needed. For example, an elephant that is used to moving only through a particular door may be hesitant to go through a different one. If the first option is unavailable for some reason, your management options could be limited significantly.

These are ways that keepers can provide reliable cues for unpredictable events. Animals recognize the concept for which keepers are asking but do not always know the way in which or when it will happen. The challenge here is to ensure that change does not happen too quickly, as individuals vary in their ability to adapt and respond. When the three elephants arrived from the Calgary Zoo, NZP keepers learned quickly that an unexpected change typically triggered escalating excitement and led to confrontations with Swarna. Rani, in particular, became overstimulated by the change, and her excitement spread to the other elephants. The keeper staff solved this problem by minimizing unexpected changes in their routine to build the elephants' confidence, after which they introduced changes slowly and incrementally. As Rani and the others became accustomed to the idea of variety in their routine, the keepers were able to make larger changes without causing social conflict. Today, the elephant team at NZP tries to change at least one thing in their routine each day. Because of the incremental steps, the elephants accept these changes without becoming too excited or frustrated.

SOCIALIZATION

As mentioned previously, socialization is a key theme throughout elephant literature and the Elephant Wellness Workshop (2016). Within our program, we highlight sociality through opportunities to interact with conspecifics, but also with care staff and visitors. The more appropriate facilities are in general (e.g., pools, shade, feeding options, etc.), the more elephants can benefit from and enjoy those features when together. If the facility provides opportunities for elephants to engage in natural behaviors such as foraging for food, bathing, and resting in the shade without having to compete for those resources, then normal social networks can develop and thrive. Those social networks and all the interactions that accompany them are inherently enriching. As facility and program managers, we must think about enrichment less because they become their own enrichment. As a visual example of this concept, simply consider the following two pictures (Figure 18). In both, elephants are enjoying a walk along the Elephant Trails trek; the main difference is that in the second picture, the elephants chose to experience the trek socially. With no other changes, the simple addition of the social element can make an experience much more enriching.

FIGURE 18: SOCIAL OPPORTUNITIES ON THE ELEPHANT TRAILS
TREK SHOWING (A) BOZIE WALKING BY HERSELF, VERSUS THE SOCIAL
CHOICE OF (B) AMBIKA AND SHANTHI WALKING TOGETHER.

TRAINING

Training is a management tool we use to care for elephants better. It improves our ability to provide for elephant health and physical needs, and requires them to employ their considerable cognitive abilities. During training interactions, an elephant is required to make choices about her response, solve problems, and master new concepts. The process can be mentally stimulating and is considered part of the enrichment program overall.

FREE CHOICE

Choice is an integral part of working in protected contact. The elephants choose constantly whether to interact with keepers. When they choose to interact with their environment, keepers attempt to keep as many options open as possible by keeping the building doors open throughout the year. They have to choose where they want to go and who they want to be near (or not). With our location in Washington, DC, there are some weather limitations, but it is a priority to provide full access as much as possible.

Throughout the workshop (Elephant Wellness Workshop, 2016), we heard a great deal about the concepts of choice and control. The ability to make decisions is an important part of animal life (e.g., Rumbaugh, Washburn, & Savage-Rumbaugh, 1989; Owen, Swaisgood, Czekala, & Lindburg, 2005; Buchanan-Smith & Badihi, 2012). Exhibit "hardware" and "software" provide numerous opportunities for elephants to make decisions, but is there a point when there is too much choice? We should celebrate choice but also be aware of its limitations. If we go back to the example of protected contact, it inherently puts the elephants in control. It requires a great deal of trust in relationships to surmount challenges in new ways, particularly when it comes to illnesses. Suppose we have an elephant who is dealing with a health challenge. She seems to be tired of injections and is exercising her freedom to choose by choosing not to cooperate. Is this a good thing? If your children could choose to get an injection at the doctor's, do you think they would do so? If a gambling addict chooses to stay at the racetrack all day and loses his rent money, was it a good choice for him? While choice is a wonderful concept, and one that is important for animals to be able to exhibit, we should be aware of the potential implications of too much choice.

Our goal with choice is to incorporate it into the social routine, access to different enclosures, etc. As we develop multigenerational herds, we are depending increasingly upon the elephants to manage themselves with respect to the elements on which they rely, keepers versus maintaining their own activities (they will fill up a lot of their own day). For that to be successful, they must have some choice and control over their life.

LONG-TERM GOALS

The hospice approach to unstable, patchwork groups, although necessary in the past due to exhibit constraints, limited collection animals, and the tendency to rely on imports rather than breeding success, is yesterday's model. We need to look for a long-term model in which we stop making the same mistakes. NZP wants to create multigenerational herds that, as they grow, can exhibit fission naturally and result in opportunities to create groups for other facilities. Our target should be to serve as a successful model of natural elephant social structures—not just to sustain the population, but also to help it thrive, behaviorally and demographically.

LESSONS LEARNED

A complex enclosure combined with an increase in group size and social opportunities can present some challenges. Providing choice through exhibit design and managing social complexity is staff-intensive. A new exhibit designed to meet the needs of animals should be the way of the future, but it is important for zoo management to understand what it means to have the resources for success. All components must be considered, from the time it takes to shift animals to determining the daily puzzle of mixing and mingling elephants. A thoughtful and planned approach to an elephant program's needs (including supplies, maintenance, and staffing) will increase the likelihood of realizing that program's goals. It is far easier to respond to issues if you have planned ahead, rather than to react after the fact.

OUTCOME

A presentation given 10 years ago, when the zoo was finalizing the exhibit design, included a statement that suits our discussion to this day: "Design assists us in managing elephants well rather than managing elephants well *despite* the exhibit." Although there were challenges and compromises along the way, NZP was able to sustain its vision to the end of construction and thereby created an exhibit focused on the physical and behavioral needs of elephants.

With respect to herd management, the journey continues. The team is focused now on merging the two herds and quantifying the effects of exhibit changes on elephant behavior and health to inform our future decisions better and contribute to evidence-based animal management. The changes we are making will be assessed with short- and long-term data. Given elephants' long lives, we should employ longitudinal studies to see effects 10-plus years in the future. NZP is excited about continuing the journey to improve elephant well-being. Now that we no longer have "elephants in the box," we have additional opportunities to engage the public with welfare-based science.

ACKNOWLEDGMENTS

We are grateful to many individuals for their help in inspiring this manuscript and facilitating its writing. We wish to thank Marie Galloway, Debbie Flinkman, Matt Chambers, Jason Gue, Becky Riley, Amanda Bobyack, Kayleigh Sullivan, Paige Babel, and Becca Spickler for their dedication to the animals at NZP and their insight about elephant behavior and NZP history; to Dennis Kelly and Dr. Brandie Smith for supporting our participation in the workshop and this book; and to Dr. Terry Maple, Valerie Segura, Dr. Joe Erwin, and the Jacksonville Zoo staff for hosting the Elephant Wellness Workshop and editing this book.

REFERENCES

Archie, E. A., Moss, C. J., & Alberts, S. C. (2006). The ties that bind: Genetic relatedness predicts the fission and fusion of social groups in wild African elephants. In *Proceedings of the Royal Society of London B: Biological Sciences, 273*, 513–522.

Association of Zoos and Aquariums (AZA). (2016). *The accreditation standards and related policies.* Silver Spring, MD: Association of Zoos and Aquariums.

Bassett, L. & Buchanan-Smith, H.M. (2007). Effects of Predictability on the Welfare of Captive Animals. *Applied Animal Behaviour Science, 102,* 223-245.

Brent L., & Weaver D. (1996). The physiological and behavioral effects of radio music on singly housed baboons. *Journal of Medical Primatology, 35,* 370–374.

Buchanan-Smith, H. M., & Badihi, I. (2012). The psychology of control: Effects of control over supplementary light on welfare of marmosets. *Applied Animal Behavior Science, 137,* 166–174.

Burghardt, G.M. (1996). Environmental Enrichment or Controlled Deprivation? In: The Well-Being of Animals in Zoo and Aquarium Sponsored Research. (Eds.: J.T. Bielitzki, J.R. Boyce, G.M. Burghardt., D.O. Schaefer). Greenbelt, MD: Scientists Center for Animal Welfare.

de Silva, S., Volker, S., & Wittemyer, G. (2016). Fission-fusion processes weaken dominance networks of female Asian elephants in a productive habitat. *Behavioral Ecology, 28,* 243–252. doi: 10.1093/beheco/arw153 de Silva, S., & Wittemyer, G. (2011). A comparison of social organization in Asian elephants and African savannah elephants. *International Journal of Primatology, 33,* 1,125–1,141.

de Waal, F. (1989). The myth of a simple relation between space and aggression in captive primates. *Zoo Biology, 8,* 141-148.

Douglas-Hamilton, I. (1998). Tracking African elephants with a global positioning system (GPS) radio collar. *Pachyderm, 25,* 81–92.

Elephant Wellness Workshop. (2016). Jacksonville Zoo, Jacksonville, Florida.

Evans, K. E., & Harris, S. (2008). Adolescence in male African elephants, *Loxodonta africana,* and the importance of sociality. *Animal Behavior, 76,* 779–787.

Fishlock, V., & Lee, P. C. (2013). Forest elephants: Fission-fusion and social arenas. *Animal Behavior, 85,* 357–363.

Fowler, M. E., & Mikota, S. K. (2006). *Biology, medicine, and surgery of elephants.* Ames, IA: Blackwell.

Greco, B. J., Meehan, C. L., Miller, L. J., Shepherdson, D. J., Morfeld, K. A., Andrews, J., Baker, A. M., Carlstead, K., & Mench, J. A. (2016). Elephant management in North American zoos: Environmental enrichment, feeding, exercise, and training. *PLOS ONE, 11*(7), e0152490. doi: 10.1371/journal.pone.0152490

Hancocks, D. (1995). Lions and tigers and bears, oh no! In B. G. Norton, M. Hutchins, E. F. Stevens, & T. L. Maple (Eds.), *Zoos, animal welfare, and wildlife conservation* (pp. 31–37). Washington, DC: Smithsonian Institution Press.

Herrelko E. S., Buchanan-Smith H. M., & Vick S.J. (2015). Perception of available space during chimpanzee introductions: Number of accessible areas is more important than enclosure size. *Zoo Biology, 34,* 397–405.

Kurt, F., & Garai, M. E. (2006). *The Asian elephant in captivity: A field study.* United Kingdom: Cambridge University Press.

Lee, P. C. (1987). Allomothering among African elephants. *Animal Behavior, 35,* 278–291.

Meehan, C. L., Mench, J. A., Carlstead, K., & Hogan, J. N. (2016). Determining connections between the daily lives of zoo elephants and their welfare: An epidemiological approach. *PLOS ONE, 11*(7), e0158124. doi: 10.1371/journal.pone.0158124

McDermott, J. & Hauser, M.D. (2007). Nonhuman primates prefer slow tempos but dislike music overall. *Cognition, 104,* 654-668.

Mingle, M. E., Eppley, T. M., Campbell, M. W., Hall, K., Horner, V., & de Waal, F. B. M. (2014). Chimpanzees prefer African and Indian music over silence. *Journal of Experimental Psychology: Animal Learning and Cognition, 40,* 502–505.

Moss, C. J. (1988). *Elephant memories: Thirteen years in the life of an elephant family.* New York, NY: William Morrow.

Moss, C. J., & Poole, J. H. (1983). Relationships and social structure of African elephants. In R. A. Hinde (Ed.), *Primate social relationships: An integrated approach* (pp. 315–325). Oxford, United Kingdom: Blackwell.

Olson, Deborah. (unknown). *Elephant husbandry resource guide*. Retrieved from https://elephantconservation.org/iefImages/2015/06/CompleteHusbandryGuide1stEdition.pdf.

Owen, M. A., Swaisgood, R. R., Czekala, N. M., & Lindburg, D. G. (2005). Enclosure choice and well-being in giant pandas: Is it all about control? *Zoo Biology, 24*, 475–481.

Prado-Oviedo, N. A., Bonaparte-Saller, M. K., Malloy, E. J., Meehan, C. L., Mench, J. A., Carlstead, K., & Brown, J. L. (2016). Evaluation of demographics and social life events of Asian (*Elephas maximus*) and African elephants (*Loxodonta africana*) in North American zoos. *PLOS ONE, 11*(7), e0154750.

Rimpley, K., & Buchanan-Smith, H. M. (2013). Reliably signaling a startling husbandry event improves welfare of zoo-housed capuchins (*Sapajus apella*). *Applied Animal Behavior Science, 147*, 205–213.

Ritvo, S.E. & MacDonald, S.E. (2016). Music as enrichment for Sumatran orangutans (*Pongo abelii*). *Journal of Zoo and Aquarium Research, 4*, 156-163.

Rosales-Ruiz, J., & Baer, D. M. (1997). Behavioral cusps: A developmental and pragmatic concept for behavioral analysis. *Journal of Applied Behavior Analysis, 30*, 533–544.

Ross S. R., Calcutt S., Schapiro S. J., & Hau J. (2011). Space use selectivity by chimpanzees and gorillas in an indoor-outdoor enclosure. *American Journal of Primatology, 73*, 197–208.

Rumbaugh, D. M., Washburn, D., & Savage-Rumbaugh, E. S. (1989). On the care of captive chimpanzees: Methods of enrichment. In E. F. Segal (Ed.), *Housing, care, and psychological wellbeing of captive and laboratory primates* (pp. 357–375). Park Ridge, NJ: Noyes.

Snowdon, C. T., Teie, D., & Savage, M. (2015). Cats prefer species-appropriate music. *Applied Animal Behavior Science, 166,* 106–111.

Sukumar, R. (1992). *The Asian elephant: Ecology and management.* United Kingdom: Cambridge University Press.

Wemmer, C. M., & Christen, C. A. (2008). *Elephants and ethics: Toward a morality of coexistence.* Baltimore, MD: Johns Hopkins University Press.

Wilson, M. L., Bloomsmith, M. A., & Maple, T. L. (2004). Stereotypic swaying and serum cortisol concentrations in three captive African elephants (*Loxodonta africana*). *Animal Welfare, 13,* 39–43.

Chapter 14

A CASE STUDY

THE REALITIES AND COMPLEXITIES OF ZOO DESIGNS FOR ELEPHANTS AT ZOO ATLANTA

Nevin Lash
Ursa International

As a zoo designer, I begin here by recommending some ground rules for designing an elephant exhibit design that focuses on wellness. To evaluate an exhibit's success, I use these 10 requirements to ensure that we provide the most we can for the wellness of the animals in our future bio-community:

1. Elephant's Choice—No human interference.

2. Exercise—Walking, reaching, swimming, hill climbing.

3. Exploration—Trunk manipulation and habitat use.

4. Proper Social Environment—Old and young (both sexes are still under discussion).

5. Safety—Keepers and elephants (inside and out).

6. Shade—In a variety of locations throughout the day and other cooling options (e.g., dusting).

7. Water—Clean water from shallow to deep, muddy water and a fresh water source.

8. Food—A variety of healthy foods, smells, textures, substrates, browse, and other forms of enrichment items to self-select.

9. Positivity—Positive human interaction and physiological considerations for both keepers and zoo visitors. Allow for significant flight distances and positive reinforcement training.

10. Vistage—A habitat with a view: long views, short views, high and low opportunities.

With these goals in mind, we can begin to judge zoo enclosures' ability to promote wellness. This chapter describes addressing these goals for wellness in exhibit design and uses a particular case study, Zoo Atlanta's new Elephant Savanna Expansion, to discuss the many influences that affect the process of exhibit design. Using this example, we discuss some of the challenges we face in building the next generation of elephant enclosures for today's zoos and sanctuaries.

My story begins with a secret meeting at Zoo Atlanta on a chilly morning in January 2014. The zoo director asked The Epsten Group and me to prepare a feasibility study of the adaptive reuse of the Cyclorama (a historic museum featuring a 360-degree painting of the burning of Atlanta during the Civil War) at the front door of the zoo. He asked, "Is there anything we can do with this thing?" You mean this massive, 4-story museum at the entry to your zoo, this amazing gift from the city? We could think immediately of several recommendations.

Secret meetings (with only top zoo executives; the elephant department was not included) continued over the next several weeks, with site visits and exploration of the 50,000-square-foot building. The Epsten Group focused on the building while I focused on the property. The museum is located in a park without any property line demarcation. However, as a private entity, the zoo has a specific property boundary, but what land should come with the Cyclorama? We knew that taking any land in Grant Park would have political implications, but new land for the zoo would expand its capability greatly.

We ended with three options: 1) minor land (50 feet) around the building, plus the entry plaza adjoining the zoo, with views into the existing zoo; 2) an additional 1.5 acres at the rear that could adjoin the existing savanna and include more hoofstock; or 3) an additional 3 acres behind the building. We thought that 3 acres would be the most that the city would permit, and with that amount of land, we certainly could create a new elephant exhibit. The zoo was at a critical juncture in its elephant program, and this acquisition would give them what they needed to build an AZA-standards program. Further, the building and its 3 acres could replace the old administration building and provide a multistory, thousand-person ballroom and events center with views into a mixed species, elephant/giraffe/rhino savanna panorama (take that, Georgia Aquarium)! We had to bring these options to city hall and let the mayor decide.

Let me return now to 1986, when CLRR, Inc. (with me as the project manager) was contracted to renovate the old Atlanta Zoo. The zoo had a terrible reputation, as it was ranked one of the 10 worst zoos in the country, and the city decided finally to do something about it. They hired a new zoo director, Dr. Terry Maple, and he selected the young firm from Philadelphia, in joint partnership with a local architect and an engineering firm in Atlanta, to develop a new master plan. With a budget of $18 million, the team set out to develop the first three phases of Zoo Atlanta to transform the zoo into a modern facility.

That story has been told many times, but I want to give the reader some context. The third phase of the renovation involved the East

African savanna animals. The zoo was able to expand approximately 6 acres into the park, and a new exhibit sequence called the Maasai Mara was born. At this time (the mammal curator was a young guy named Tony Vecchio), CLRR did their best to create a set of exhibits for elephants, black rhino, giraffe, zebra, antelope, and ground birds across from a lion exhibit, as well as some visitor services facilities on a 9-acre footprint for approximately $6.5 million. The exhibit opened in 1998 and was a great success, completing the initial three-phase transformation of Zoo Atlanta and receiving AAZPA Exhibit Awards for both the West African gorilla exhibit and the East African savanna.

Fast-forward to July 2014, when the mayor announced, "The Cyclorama (painting and locomotive) will move to the Atlanta History Center, and by the way, we are giving the zoo the empty building plus 3 acres of parkland and 2 acres of building site." The zoo braced for a neighborhood rebellion. However, it transpired that the neighborhood was so concerned about moving its cherished Civil War painting out of Grant Park that the residents scarcely noticed that the zoo was taking over precious parkland. Community voices directed to the zoo were more positive in nature—that it could use more land and would put the building to good use. Previously, the zoo had been chastised for encroaching into the park, but now it was acceptable because there were worse things happening! Our Civil War monument is moving to the HISTORY center! The ruckus settled down quickly as the story emerged that, with its eye on the painting, the history center had a $10-million building designed and ready for construction and that the zoo had a new vision for the great Cyclorama.

The design team received permission to design an ultimate new elephant enclosure (see Figure 1) on a site that we could connect easily to both the giraffe/hoofstock areas and the rhino habitat to create an expanded savanna exhibit. This expansion was unforeseen during the last master planning effort and would change the current master plan significantly (i.e., elephants were planned at the opposite corner of the zoo grounds). Changing the master plan was not a problem for the design team, but in Atlanta, you have to have approval from the Urban Design

Commission to obtain a building permit, which requires the project to be based on an approved master plan.

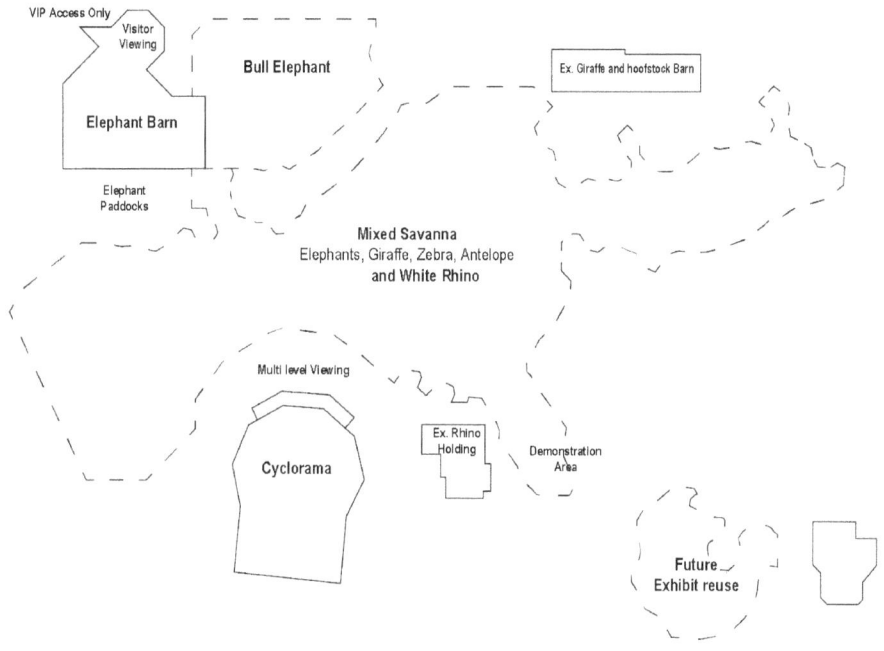

FIGURE 1: VISITOR CIRCULATION DIAGRAM, EARLY CONCEPT SITE EXHIBIT PLAN.

The exhibit that we designed combined the existing giraffe savanna with the expansion land to create one, large (4-acre), multispecies (giraffe, zebra, waterbuck, warthog, white rhino, and cow elephant) lower terrace habitat with a smaller bull yard and paddocks in the upper terrace that could be interconnected. The proposed barn sat on the perimeter service road in the expanded area, and visitors would be able to access it only in VIP programs. Because we were working in "secret" during the early stages, we knew the day would come when the zoo's animal department and our design team would need to reach consensus, at which time our design either would sink or float. I knew I was asking a lot to have a mixed-species elephant/giraffe habitat, but we had to dream big.

As the design progressed, we finally met openly with zoo staff. The zoo's elephant manager, Nate Elgart, had done his homework. He walked into our first meeting and announced, "There will be no giraffes in the new elephant habitat, and we need a 19,000-square-foot elephant barn for seven elephants, including two bulls." This was 9,000 square feet more than I had budgeted during our secret meetings, so we began fairly far apart. The design could not move forward until we reached consensus on an acceptable barn size. The design team had built-in fees to tour other recent elephant exhibits to see how other zoos approached the next generation of elephant exhibits. We decided to visit Dallas, Cleveland, North Carolina, and Birmingham. With that tour agreed upon, Nate Epsten, architect Pete Choquette, and I set out to look at those recent elephant facilities to prepare our redesign of the project.

We visited Dallas Zoo's Giants of the Savanna project first, which opened in 2010. Here, we would see whether it was feasible to combine elephants with giraffes. We saw immediately the value of a lightweight metal barn for cost efficiency, with roll-up doors, a large herd room, bollard walls, large sand floors, Big Ass fans, radiant heaters, and a 5-ton bridge crane. In the habitat, we learned about overhead feeding stations for public "demonstration" areas as well as keeper lookout stations in the rear of the habitat. We also spoke to elephant manager Karen Gibson about combining giraffes and elephants. Unfortunately, we were not there during their one-day-a-week mixed savanna, but just talking to Karen, I knew it was possible for us to attempt it. It would take time and effort, but the payoff would be extraordinary.

Dallas Zoo's barn is not on view to the public. At the back of the habitat, they use heavy cable barriers with an added mesh layer to keep the antelope in the enclosure. On the visitor side, there is an amazing 180-degree wide-open view looking out over a huge water moat and pool. Alan Roocroft (the elephant whisperer from San Diego) was sitting under a tree in a director's chair, apparently the man behind the magic at Dallas. Alan was training several staff about their new-ish exhibit and having good results. We learned that to succeed in creating these more complex exhibits, money needs to be set aside for staff training, and that training takes time.

That night, we flew to Cleveland and landed in a snowstorm. We had an in-depth tour and visited their large elephant collection when everyone was inside. At Cleveland, we learned that you can create an "on grade crossing" of elephants and visitors, but you may have to tolerate some pathway bottlenecks (those are exciting to watch!), and you better invest in high-end technology and premium design/construction to achieve it—which they did and do. However, Nate was uncomfortable with the complex management this visitor crossing required each time an elephant shifted.

Inside the elephant barn, which allows visitor viewing, we looked forward to seeing the way they managed the cable barrier at the front of the stalls. To our surprise, the cables are electrified. This reduces the footprint in the stalls further, and they are cramped already. Nate had strong feelings about potentially allowing unescorted visitors to touch an elephant inside and did not believe the hotwire solution was acceptable. That was a clear veto of cable barriers in our barn. The rest of their barn is remarkable for its extra-large everything—from the containment, extensive air handling, hotwire, and filtration systems to the custom-built drinkers and huge hayloft.

There was a lot to like about Cleveland's outdoor elephant habitats. It could have been the nice coating of new snow or the pale-yellow cable barriers that outline a gracious pair of open habitats. We could imagine that Cleveland's elephants lead a pretty comfortable life. The habitat has large pools for open viewing, large rectangular shade structures, several large existing oaks (protected with hotwire) to provide afternoon shade, as well as themed visitor huts and trellises. We learned that the Cleveland model of "Elephant Crossing" was feasible, but not for us, as it would not sufficiently improve our elephants' well-being to counteract the increase in staff efforts and potential safety issues.

A week or two later, we visited North Carolina, where Guy Lichty was kind enough to give us the VIP treatment. After studying the plan layout, we went out to the site to view the reality, which is even more impressive. Their 7-acre habitat (10 overall) has expansive viewing with a combination dry moat and a huge pool with an underwater barrier on the visitor side, and an 8-foot cable rail in the distant edge of forest. There are several large enrichment structures to scratch on and certainly enough room to roam. At the elephant visitor center (themed as a helicopter hangar), we were able to observe the manner in which they shift their elephants from habitat to habitat, which is a demonstration for guests. The shift appears to provide an easy and safe interface among the elephants, keepers, and guests. It also provides considerable flexibility and uses for individuals or groups of elephants.

Then we drove back to the barn, to which there is no public visitor access. One of Guy's first comments to me was, "We wish we had visitor access to the barn for a variety of reasons, like winter, rain, and educational value." North Carolina Zoo typically does not have behind-the-scenes viewing, period. However, should we give access at Zoo Atlanta or not? We were hearing mixed opinions.

The hilly terrain, wide-open views, hidden moats, and distant viewing make this place an outstanding example of seamless habitat design. North Carolina Zoo staff were engaged fully in not only the design process but in fabricating the containment systems, purchasing and installing landscape plant material and soils, and developing and coordinating the production of all graphics and interactive material. They created an interactive game at the Elephant Exhibit so children can learn about life on the savanna in an enjoyable way. At North Carolina, we learned that we do not have nearly enough acreage to do it right—10 acres minimum—instead, we have less than 5. However, we would have the opportunity to improve conditions for the elephants at Zoo Atlanta greatly.

Next on our list was Birmingham, just two hours from Atlanta. Their elephant exhibit is a nice combination of North Carolina and Dallas, with an open viewing habitat and the most perfect four-bull stall barn we had seen. Birmingham has huge sand stalls with a transfer chute of bollards in front of them. This allows direct elephant/keeper access and circulation. It also has several large outdoor paddocks and one isolation suite. Their habitat is spacious (4 acres), with all-natural vegetation at some level of destruction. The soils are muddy, deadfall is strewn everywhere, and the bulls love it. There was a great deal of activity, and the elephants looked wonderful. As we learned at the Jacksonville Elephant Conference, bulls typically live on their own or with other bulls in the wild. Having an all-bull zoo habitat makes the most sense as a strategy, rather than each zoo building to accommodate both cows and bulls. Currently, zoos force the bulls to live with cows all their lives, with occasional separations by keepers. Does this promote elephant wellness?

Our multizoo site visits were eye-opening for all of us. They offered us, as a design team, opportunities to learn from one another and to increase our understanding of the scope of our project. We saw many forms that our program could take by seeing the similarities and differences in each zoo and the way their components were consistent with our team's philosophy of elephant exhibit design.

Now our design process could really begin in earnest. We knew what we believed were the best parts of each design. Between Nate, his staff, and me, we discussed barn layouts from the sublime to the ridiculous and, with refinements, reached a final agreement. Ultimately, we chose a perfect elephant barn that is 19,000 square feet and can accommodate up to seven (2.5) elephants. It includes a large herd room, three cow stalls (all with keeper access halls), and two large bull stalls. The bull stalls are connected across the hall from the cow stalls, all with access to outdoor paddocks. It also includes an elephant restraint device (ERD) chute that leads to two interconnected habitats. Our hillside location allowed us to terrace into the hill, with a 10-foot-high clay bank separation between the two habitats. This provides a very deep exhibit vista and has the flexibility to move elephants to either level and giraffe and rhino into either habitat as well (Figure 2). Visitors will be able to see elephants from the Cyclorama viewing terraces, and we designed a special visitor path that wraps around just one side of the habitat as well to give the elephants maximum acreage. This provides long, panoramic views into different portions of the savanna habitat. We all agreed to allow visitors access to the first floor of the elephant barn, just in case the elephants are inside. The view in the barn will be from the ground (a "respectful" orientation to promote wellness). At that point, visitors will have to turn around and return the way they came. While not ideal, it was acceptable, and this cul-de-sac path will allow the visitors to review what they have seen from a new perspective, and help them absorb what they learned about the state of elephants in the wild. If we had chosen a visitor loop, it would have had to cross elephant pathways and, therefore, would have required either an elaborate bridge (like Denver), on-grade crossings (like Cleveland), or a wooden ramp (like Oklahoma City). This would constitute an additional scope of work never envisioned in the concept plan and, thus, not in our budget (or so I thought).

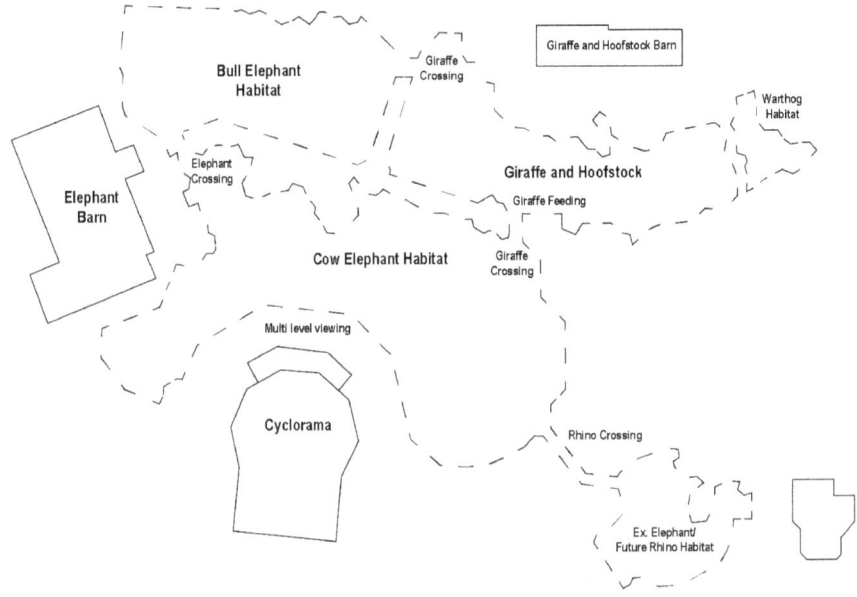

FIGURE 2: VISITOR CIRCULATION DIAGRAM, SCHEMATIC SITE EXHIBIT PLAN.

The zoo staff loved the plan and the approach and encouraged us along. We had a visitor-accessible barn with an elephant exhibit easy to manage without visitor interference. Our final schematic design exhibit plan was approved, and we moved on to design development.

However, there still was that nagging master plan. The zoo brought back their Master Planner, Torre Design Consortium to update their master plan to include our elephant exhibit layout. When they saw the visitor circulation, they noted the cul-de-sac visitor path and convinced Zoo Atlanta's CEO that there needed to be a loop, as loops keep visitors moving and increase their stays. They argued that the length of the path equaled more revenue. They did not consider elephant management or the budget: revenue was the highest priority, not wellness.

From that time forward, the loop was to be. I returned from a week vacation in Belize to hear, "We want to add a visitor loop, and it goes right through the middle of the exhibit!" Our schematic design had been complete, but now we had to look at the implications of this little squiggly magic marker line cutting through the middle of the exhibit (Figure 3). Having some respect for budget and the laws of physics, my first response was simply to say "NO." However, as we "serve at the pleasure" of our clients, I had to demonstrate a better option.

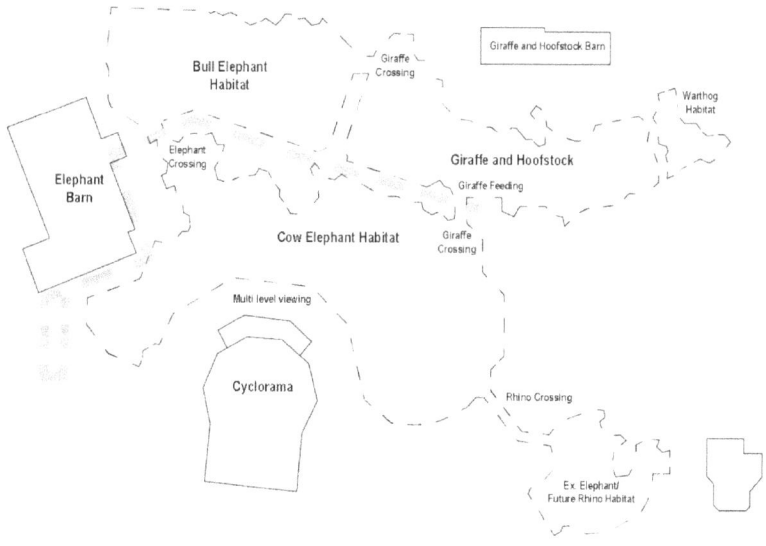

FIGURE 3: VISITOR CIRCULATION DIAGRAM, SCHEMATIC SITE EXHIBIT PLAN WITH POSSIBLE LOOP.

To accomplish the loop, we would keep visitors on the ground (there was neither sufficient room nor budget for the ramps or boardwalk) and create a Cleveland-style crossing at the cow access point. The elephants would have to be monitored to allow one group of animals to move from one habitat to another or to the barn. Visitors would have to stop and bottleneck on the path until all people doors were closed, all animal doors were opened, the animals moved through, the animal doors were closed, and the people doors opened, after which they could move along. This was unacceptable for many reasons—in addition to safety and logistics, it would restrict both animal and visitor movements; would put people in the middle of the panorama view from the Cyclorama; would take potential animal habitat; and, of course, it would cost more. Nevertheless, the most significant issue was that keepers would decide for the elephants when it was time to move and where, rather than keeping the entire system open and allowing the elephants to choose where to go. Isn't that the most important ground rule for wellness—giving the elephants the choice and freedom to explore, exercise, and experience new places on *their* terms (or to the extent we can allow)?

There was another possibility. We could cross over the top of the elephants as they do at Oklahoma City Zoo. To accomplish that, we would have to take the people up an 8% ramp to a minimum height of 24 feet above the elephant floor before they crossed over the top; this resulted in a 450 LF ramp just to reach the top. We would need room for the ramps to extend around the entire exhibit; we were not going to allow visitors through the middle of the habitat—that simply was *too* much.

The "revised" plan (Figure 4) resulted in an exhibit with:

1. Visitors (with their potential airborne debris) standing above the shifting and paddock spaces below;

2. Encircled elephants, with visitors at different levels while in the habitat;

3. Reduced elephant area because of additional walkways with 15-foot setbacks and a new central viewing node (a 10% total reduction in elephant area);

4. Cross-viewing and back-of-house viewing from new ramps, walkways, and Cyclorama decks;

5. Reduced landscape buffer as a backdrop;

6. Additional construction costs involved in building a massive ramp and a lengthy path that required additional interpretive and activity areas, restrooms, snacks, and shade.

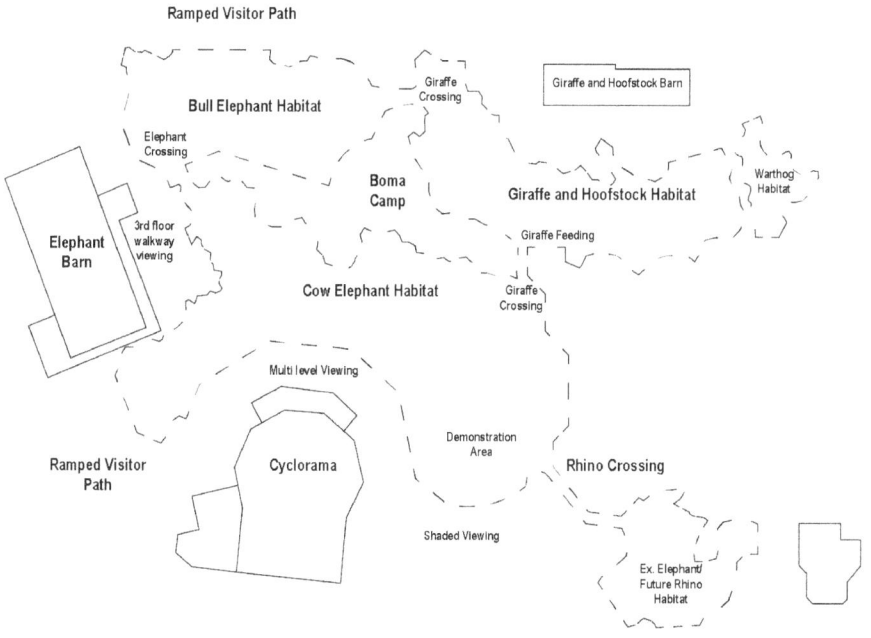

FIGURE 4: VISITOR CIRCULATION DIAGRAM, DESIGN SITE EXHIBIT PLAN WITH LOOP.

Once we embarked on this path, the design evolved to include a revised barn that would move visitors from the first to the third floor, where we would provide an indoor viewing/interpretive room that overlooked the massive nave of the elephant cathedral. Visitors also would be able to walk along an outdoor balcony that runs halfway down the length of the barn to a covered overlook of the habitats.

Then the visitor ramp down would continue to the far perimeter of the exhibit (along the service road), where it would provide excellent views of the bull's yard and his waterfall and pond. Farther down the trail, the visitors would pass by a transfer where giraffes enter the bull yard through a series of gates and chutes. Finally, they would arrive at a "boma camp" located in the center of the exhibit, with views of both elephant habitats and the giraffe/hoofstock savanna exhibit. This central node is themed as a traditional African hut with secondary huts for restrooms and beverages in a camp setting.

From there, visitors would return to the "old zoo" at a very busy, expanded giraffe feeding platform with more views of the elephant pool and cow demonstration area. This route resulted in an 1800 LF loop. It would be a long walk for our guests, perhaps too long. It also would be a very different type of exhibit for Zoo Atlanta, with continuous viewing along all sides and little area left for buffer landscapes. We still maintained a 15-foot setback from the elephants that would allow the keepers many training opportunities and access to the elephants. This was a much bigger visitor experience than we had agreed on originally, and having visitors encircling the elephants and viewing them from the third-floor balcony of the barn is not ideal for elephant wellness.

We also still needed to address the budget. We commissioned another construction cost estimate only to find that, as expected, the additional "loop concept" added another $3 million-plus in costs! Even with extensive value-engineering efforts, we could not offer a solution that maintained the loop and the budget at the same time: something had to give.

In summary, how does wellness affect exhibit design? When is it necessary to change one's business plan (need for a loop path) in exchange for animal wellness considerations (visitors on all sides, from all levels)? Did Zoo Atlanta make the right choice in assessing a revised exhibit design? Does the psychological issue of being surrounded contribute to a feeling of distress? The Manitoba standards for polar bears include a requirement to limit visitor/human interfaces to 180 degrees—should we consider that for elephants as well?

POSTSCRIPT
(AFTER THE ELEPHANT CONFERENCE)

At the time of this lecture, we were waiting for the zoo to decide which way it would proceed. Several Zoo Atlanta staff attended the Jacksonville elephant conference, and I like to think that my presentation represented the original concept plan as clearly beneficial to elephant wellness, and the loop concept less so.

Ultimately, the zoo reached its comfort level with the project's budget, and we were directed to remove the ramps, pathways, third-floor elephant viewing, and maintain the project budget (i.e., return to the schematic layout approved). I believe that this was the right choice, even if the zoo did not see its merits initially. I feel that while the ramped overlook experience would offer something that the zoo does not currently, the operational, capital, and potential safety costs were too great for this project. We will never know whether the extensive path system and encircling elephants at this scale would have been detrimental to their well-being.

Since given permission to pursue the original cul-de-sac path system, we decided to revisit the concept and attempt to improve visitor stay-time by adding additional species and experiences. Originally, meerkats were eliminated from the project, but now they were targeted immediately as a potential new and popular exhibit. Hooded vultures could be added at limited costs and contribute to the species survival plan for this critically endangered species. The new plan now provides a generous visitor area in front of the barn that offers a very long view of the habitats. Additional features include an overhead feeder and interactive discovery wall where the visitors will be able to observe, close-up, several behaviors that make elephant exhibits some of the most popular at any zoo. Now the visitors will not just come to the barn and turn around; instead, they will be able to rest, view, get a beverage, and learn directly from zookeepers in and outside the barn––all in an African-themed "cool spot" filled with unique savanna animals, and ones that need our help. Isn't this part of the definition of wellness—to care for animals and save them from extinction? We now have a new direction in which wellness is the priority.

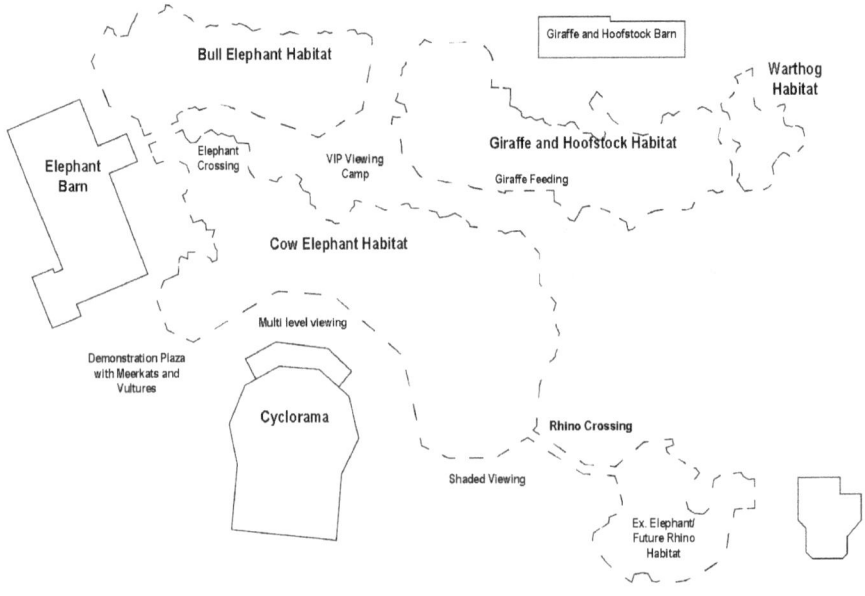

FIGURE 5: VISITOR CIRCULATION DIAGRAM, FINAL SITE EXHIBIT PLAN.

The new Zoo Atlanta savanna expansion addresses my list of the behaviors necessary for a successful exhibit that promotes wellness:

1. Animal's choice without human interference

2. We will be able to open the entire elephant area to the herd without monitoring gates or close down to separate enclosures with enough buffer between groups. It was not the best scheme for animal choices as we separated the giraffe, but it remains possible to add white rhino.

3. Exercise—walking, reaching, swimming, hill climbing

4. There will be many exercise opportunities in the 450-foot-wide habitat with reaching, swimming, and hill climbing available—the hillside location ensures this.

5. Exploration—trunk manipulation and habitat use

6. We will begin with one interactive wall with treats at the demonstration wall and an overhead feeder. Over time, more can be added. The 15-foot-wide setback will give the keepers the ability to enrich the elephants' habitat with novel items at will.

7. Proper social environment—both sexes and old and young

8. The herd will be formed with two females who have lived their lives together and two additional females and a male who have lived together forever as well. Bringing that group together will be the key to a harmonious social group. The outstanding question remains: are Bull Elephants appropriate for 'proper social environment' in a Zoo exhibit?

9. Safety—keepers and elephants (inside and out)

10. We have a 15-foot-wide setback from the perimeter fencing—and with a 7-foot trunk, the keepers will have 5 feet of safe area

around the entire enclosure in which to interact with the elephants in the habitat, and the same in the barn and paddocks.

11. Shade—from a variety of sources throughout the day and other cooling (e.g., dusting)

12. At this time, I am relying on existing vegetation to provide shade areas in the habitat and also large fabric umbrella "trees." There will be sand for dusting.

13. Water—clean water from shallow to deep, muddy water and a fresh water source

14. There will be two waterfalls, one of which flows along the front of the exhibit as a shallow stream to explore and ends in a large central pool, while the other is simply a soaking pond with a waterfall.

15. Food—a variety of healthy foods, smells, textures, substrates, and browse

16. The elephants will be on a variety of substrates such as grass, sand, and clay, as well as have a great deal of deadfall to play with and eat.

17. Positive human interaction and physiological considerations for both keepers and zoo visitors.

18. We will have positive reinforcement training around the exhibit perimeter and provide excellent flight distances to facilitate the training.

19. We will maintain a lower viewing aspect from the visitors to the elephants, except at the Cyclorama, which has an overview.

20. There will be no visitor and elephant crossings, only a visitor/ rhino crossing, which will offer periodic and experimental opportunities for interspecies interactions in safe conditions.

21. Keepers will have an overlook balcony and office area in the barn to observe the elephants (as well as future cameras).

22. A habitat with a view: long views, short views, high and low opportunities.

23. This is a very long exhibit (for Zoo Atlanta), and when both habitats are joined, there will be extensive opportunities for interesting viewpoints. The terraced exhibit and the Cyclorama's multi-floor balconies will offer a variety of observation points for both elephants and visitors.

The project, at the time of the publication of this paper, is on the brink of construction. We have a construction manager who is in charge of the budget and will build the exhibit for a guaranteed maximum price. In order to match budget and scope, we have had a make additional "value engineering" decisions (many were not accepted by the zoo's animal department), and additional money will have to be raised to complete the project as it should be. I think we have maintained a focus on animal wellness with these final decisions and created a framework that can be improved over time. Soon, the zoo will be able to begin that exciting process of taking their elephant program into the twenty-first century with a facility that is designed to promote their natural behavior with environmental enrichment opportunities for the elephants and safe animal handling for the keeper staff to interact and care for these great beasts. We are proud of the way we fought the good fight to make a home for elephants where they will thrive in Atlanta.

Chapter 15

CREATING A GIRAFFIC PARK

Design and Management Ideas for Optimal Giraffe Wellness

Meredith J. Bashaw
Franklin & Marshall College

CREATING A GIRAFFIC PARK

The goal of this chapter is to apply the ideas about elephant exhibits and management discussed elsewhere in this volume to another species of charismatic megafauna, the giraffe. Captive giraffes and elephants present some of the same challenges for zoos, although they also differ in important respects. Although I will focus here on giraffes, I hope to present an approach to exhibit design and animal management that starts from the animals' point of view and should be applicable broadly across species. This empirical approach includes three main facets for improving wellness: considering the animal's sensory experience, promoting healthy species-typical behavior, and applying the psychological constructs of predictability and control.

GIRAFFES' SENSORY WORLD

The "umwelt" is a term used to describe the world as perceived by a particular individual or species (von Uexküll, 1934). Because each species' sensory organs and nervous system are adapted to help it meet a specific

288

set of ecological challenges, no two species will perceive the same environment in quite the same way. As humans, we are terrestrial, bipedal creatures with trichromatic color vision. As such, we tend to naturally focus on how an environment looks and what it affords us. What can we see? How well can we move around the space on the ground? Is it too hot, loud, or smelly for us to be comfortable? You can imagine that the same environment would "look" very different to an individual of another species. For example, some birds see and rely on UV cues to mate and find food (Aidala et al., 2012). UV markings are beyond our visual range, so two birds that look identical to the human eye might look quite different to conspecifics. Our focus on how humans perceive the environment sometimes leads us to build animal habitats that are missing key features. For example, 20 years ago, indoor bird habitats rarely included a UV light source. However, research showed that some birds preferred portions of their exhibits with access to UV light and altered their social behavior in its presence (Ross, Gillespie, Hopper, Bloomsmith, & Maple, 2013), so now UV light provision is much more common. To take this a step further, how might an animal that does not rely on vision as its primary sense, such as a bat, perceive the same exhibit? Whether it is possible for us to imagine what it is like to be a bat has provoked great argument among philosophers (e.g., Nagel, 1974), but we can agree that to house a bat well, we would need to pay particular attention to aspects of the auditory environment (ambient noise levels, sound reflectance) that would be much less important for housing a human.

Several people at the Wellness for Elephants conference have described an elephant as a nose attached to two ears, emphasizing the importance of olfaction and audition in an elephant's life history. Giraffes' umwelt is still very much a mystery, as there are surprisingly few published data on their sensory capabilities. However, what we know suggests that giraffes are unlikely to perceive the world—or their exhibit—the same way as a human or an elephant.

VISION

In contrast to elephants' reliance on audition and olfaction, vision appears to be the sense most important to giraffes in predator avoidance and social behavior (Bercovitch & Deacon, 2015). Giraffes have been referred to as the watchtowers of the Serengeti. The structure of their retina suggests that they have excellent long-range visual acuity, and wild giraffes have been observed to respond to stimuli as far as 2 kilometers away (Coimbra, Hart, Collin, & Manger, 2013; Dagg, 2014; Mitchell, Roberts, Van Sittert, & Skinner, 2013). Males use visual signals to communicate dominance and submission, but these vary with distance (Seeber, Ciofolo, & Ganswindt, 2012). At a long distance, a dominant male will hold his head high and adopt a very vertical posture, while a subordinate will lower his head until his neck is parallel with the ground. At closer range, dominant males drop their heads to nearly horizontal and walk directly toward a subordinate, who will adopt a nose-up posture, lifting his head until it is nearly vertical. Cameron and duToit (2005) found that having a large adult male nearby changed the frequency of vigilance behavior by both female and other male giraffes, suggesting that giraffes monitor conspecifics for social signals visually.

Like giraffes, polar bears live in a natural habitat where they can see long distances and use visual social signals. A multi-institutional study found that captive polar bears with a view out of their exhibit paced less often (Shepherdson, Lewis, Carlstead, Bauman, & Perrin, 2013). Anecdotal observations at several zoos have suggested that when giraffes have a solid fence or wall on one side of their exhibit, individuals that are tall enough regularly devote time to looking over the wall. If long-range vision and visual monitoring are important to giraffes, then the ideal exhibit might allow them views out of the exhibit, preferably over a long distance and in many directions.

In addition to reducing visibility, captive housing typically reduces freedom of movement. Being able to escape from visual contact with conspecifics can provide a valuable opportunity to reduce conflict (Laule, 2003). Providing visual barriers, access to a barn during the day, or other retreat spaces may facilitate giraffe well-being by giving individuals the chance to avoid potentially stressful interactions with other giraffe, people, or external stimuli.

OLFACTION AND TASTE

Like elephants, giraffes use chemical cues to communicate over short distances. For example, giraffes use the typical ungulate ritual of anogenital investigation, urine testing, and flehmen to assess female reproductive state (Seeber et al., 2012). Adult males use regular olfactory investigation to identify fertile females and confine their mating activity with females to the short peri-ovulatory window (Bercovitch, Bashaw, & del Castillo, 2006). In sheep and goats, ovulation can be induced by exposure to olfactory cues present on the male's pelt (Gelez & Fabre-Nys, 2004). The pelts of adult giraffes, particularly males, have a strong aroma attributable to compounds secreted by their skin. These chemicals repel parasites and have been postulated to attract females, perhaps as an honest signal of health and fitness (Wood & Weldon, 2002). Giraffes often investigate one another by sniffing or licking (Bashaw, Bloomsmith, Maple, & Bercovitch, 2007), giving them an opportunity to detect chemicals on one another's skin. However, the effects of exposure to male scent on ovulation in female giraffes has not yet been studied. Unlike many other species for which scent is an important social signal, there is no evidence to date that giraffes scent mark or respond to scents other giraffes have deposited on objects in the environment.

Morgan and Tromborg (2007) provided an excellent review of the potential interactions between captive management and olfactory communication. For animals such as giraffes that use scent to communicate, cleaning products with strong odors can mask such cues or be overstimulating and should be avoided. Although cleaning can be stressful or stimulating for animals that use scent to identify their territories, giraffes should not be affected in this way because they have not been reported to engage in territorial marking. On the contrary, giraffes produce a prodigious amount of waste, so infrequent cleaning (especially of indoor enclosures) can result in the rapid accumulation of strong waste odors that may be aversive or affect olfactory communication. Possible effects of cleaning frequency on giraffe welfare have not yet been studied.

Giraffes' interest in scent can also be harnessed for enrichment. Fay and Miller (2015) found that the addition of novel scents to unused portions of a large giraffe yard at the San Diego Zoo's Safari Park increased time spent in the location closest to scent dispersal and decreased the amount of time animals spent standing and resting. Rose scent was preferred overall, although one giraffe liked banana and another orange. In the absence of deliberate enrichment, giraffes tend to spend most of their time in areas of their exhibit that provide food or shade (personal observation), so scent enrichment has the potential to create a more stimulating captive environment for giraffes. Captive giraffes will also take advantage of opportunities to engage in object manipulation and exploration. Providing novel manipulable objects (for example, Kong toys or small objects nested inside larger ones) and rotating them regularly can give giraffes additional opportunities to explore. Enclosure designs with numerous attachment points at different heights and locations will support the addition and rotation of toys, feeders, and other stimuli, as well as allowing the enclosure to keep pace with new developments in enrichment.

SOUND

Unlike elephants, giraffes do not appear to use sound as a primary way to stay in contact with conspecifics. However, they do produce and attend to sound cues. Giraffes produce a variety of relatively quiet sounds, including bursts, snorts, grunts, and night humming (Baotic, Sicks, & Stoeger, 2015). They also have been reported to produce infrasonic vocalizations (von Muggenthaler, 2013), although attempts to record these vocalizations in controlled conditions (Baotic et al., 2015) and to detect giraffes' responses to playbacks of infrasonic vocalizations (Bashaw,

2003) have been unsuccessful. The function of these sounds is unclear, but most scientists have assumed they are used to communicate. Brief, loud, and/or novel environmental sounds are a source of stress in captive animals (Morgan & Tromborg, 2007) and are likely to cause increased vigilance in giraffes.

The savanna is a quiet habitat, with ambient noise levels typically between 20 and 36 decibels, virtually all of which are attributable to wind and rustling vegetation (Waser & Brown, 1986). For comparison, a quiet suburban household typically has background noise that measures approximately 50 decibels (Branch & Beland, 1970). Zoos, on the other hand, are fairly loud environments. Tromborg and Coss (1995) recorded ambient noise in the San Francisco and Sacramento Zoos and found its usual range was 62 to 72 decibels, comparable to a running vacuum cleaner. Noise varied as a function of the number of guests and their conversation intensity, the presence and operation of machinery (including climate control systems and landscaping equipment), water features, and transportation systems both inside and outside the zoo. A typical human conversation occurs at 50 to 60 decibels at a 1-meter distance, depending on the level of background noise. As you have probably experienced, carrying on a normal human conversation during a typical day at the zoo can sometimes be difficult. Documented giraffe noises range from 35 to 40 decibels at a 1-meter distance (Baotic et al., 2015), so ambient noise could interfere with giraffes' auditory communication in a zoo.

Some sources of noise, such as transportation systems and proximity to urban areas, are out of a zoo's control. However, zoos can work to reduce ambient noise within their giraffe facilities in a number of ways. In designing on- and off-exhibit areas, avoid sound-reflective surfaces such as smooth block walls in favor of textured surfaces that dampen sounds (Morgan & Tromborg, 2007). Sound-absorbing or -reflecting barriers can be placed between giraffe housing and major guest or vehicle pathways; these need not be industrial in appearance, as many companies now manufacture sound-absorbent art panels. Even the simple addition of a layer of shade cloth or plywood between noisy guest areas (playgrounds, eating areas, major pathways) and giraffe housing can reduce noise.

Disney's Animal Kingdom is on the leading edge of this effort and is taking proactive measures to counteract the possible detrimental effects of noise on their animals (Soltis, this volume). For example, staff are measuring sound levels and identifying or building sound sanctuaries within their exhibits, quiet spots to which animals can retreat if they wish to avoid environmental noise. In addition, Disney is requesting that events such as fireworks and parades begin with lower-volume sounds that animals can use as a cue to retreat before the event reaches full volume. These approaches are particularly promising ways to improve welfare because they give animals a choice to avoid loud noise if or when they find it aversive. The effects of these management changes are currently under study.

PROMOTING HEALTHY AND SPECIES-TYPICAL BEHAVIOR

Animal welfare typically is conceptualized in one or more of three ways: the presence of positive emotions, the health and productivity of the animal, or the opportunity to perform natural behaviors (Fraser, 2009). At the most basic level, this requires captive environments to support maintenance behaviors such as sleep, thermoregulation, feeding, and social interaction.

SLEEP AND OTHER NIGHTTIME BEHAVIORS

Sleep patterns in captive adult Asian elephants (Tobler, 1992) and giraffes are comparable, with giraffes sleeping and average of 4.6 hours per day, sometimes while standing and sometimes while lying (Dagg, 2014; Tobler & Schwierin, 1996). Unlike elephants, it is easy to identify when giraffes engage in paradoxical (including REM) sleep; this type of sleep only occurs when giraffe are lying fully recumbent with the head on the flank or ground. Giraffes average 27 minutes per day of REM sleep, less than 10 minutes per day for adults and up to 63 minutes per day for juveniles (Sicks, 2012). Langman's (1977) observations of wild Cape giraffes concurred that calves engage in the most recumbent sleep but suggested that juveniles (1.5 to three years of age) engage least often in recumbent sleep. Studies of sleep in captive adult giraffes have revealed individual differences in the number of bouts of lying per night, with some individuals lying only once per night and others two or three times (Sicks, 2012; Tarou et al., 2001; Tobler & Schwierin, 1996). For example, Tobler and Schwierin observed two peaks in lying behavior, one from 8:00 p.m. to 2:00 a.m., and another from 4:00 a.m. to 7:00 a.m. Sicks (2016) has suggested that the amount of recumbent sleep can be used as an indicator of stress.

For giraffes, substrates are likely to be particularly critical for comfortable sleeping. Given their high center of gravity and long legs, the process of lying down and standing up is cumbersome (for illustration, see Tobler & Schwierin, 1996). Giraffes also are most vulnerable to predators while obtaining paradoxical sleep (Tobler & Schwierin, 1996; Sicks, 2012). Captive giraffes engage in less of this type of sleep following stressful events (Sicks, 2016), which may be of concern because long-term deprivation of REM sleep has negative consequences for health and learning (Siegel, 2005).

Studies of cattle show they prefer to lie down on substrates that are softer but also increase friction and resistance to slipping (Bargai & Cohen, 1992; Van der Tol et al., 2005). For example, cattle walk and sleep better on sand or rubber mats than on concrete (Telezhenko & Bergsten, 2005), with yielding rubber mats preferred to sand (Norring, Manninen, De Passille, Rushen, & Saloniemi, 2010). The provision of additional bedding (for example, hay) overtop the mats increases lying time further (Tucker, Weary, Von Keyserlingk, & Beauchemin, 2009), indicating greater comfort. Hummel et al. (2006) noted similarities between the hoof problems and arthritis developed by captive giraffes and those developed by dairy cattle, and suggested that softer substrates should be used in giraffe barns. The effects of various substrates on giraffe behavior should be assessed experimentally, but in the absence of such data, interlocking mats or rubberized flooring is advisable, with bedding provided on top of such flooring when possible. An added bonus is that these types of flooring also absorb sound.

If giraffes sleep less than five hours per night, it is reasonable to wonder what else they do during the night. In 2001, using four infrared illuminators and the light of a 7-watt red Christmas tree bulb, observers at Zoo Atlanta recorded the overnight activity of two adult female giraffes (Tarou et al., 2001). We found that in addition to sleeping, much of the night was spent in oral behaviors: feeding, ruminating, and repetitive licking of nonfood objects. During the course of the night, these giraffes engaged in fairly consistent rates of repetitive licking of nonfood objects (26% of scans) and ruminating (4% of scans). Despite continuous access to food throughout the night, they showed two peaks of feeding behavior, one just after entering the barn and another just before the care staff arrived in the morning. Duggan, Burn, and Clauss (2016) observed six giraffes at two UK zoos and found similarly high rates of repetitive licking and tongue play overnight compared to during the day. Studies of and suggestions for reducing rates of repetitive licking and tongue play are described in more detail in the section on feeding.

Historically, zoo enclosure design has focused on creating attractive, functional, interactive exhibit spaces, but animals are often in these spaces for less than half of the day. For example, according to a 2000 survey of AZA-accredited zoos that house giraffes and okapi, the average giraffe spent approximately 12.7 hours per day indoors (Bashaw, Tarou, Maki, & Maple, 2001). According to a 2004 survey of European zoos, giraffes spent a median of 14 hours per day indoors in the summer and 20 hours per day in the winter (Hummel, Zimmerman et al., 2006). Many zoos in northern latitudes need to house giraffes indoors continuously in the winter to keep them warm (a discussion of thermoregulation in giraffes follows). Like many animals, giraffes are active for much of the night. Where possible, giraffes should be allowed access to their outdoor enclosures overnight, and more attention should be paid to increasing housing complexity and the availability of behavioral options for giraffes when they must be housed indoors. Providing food that requires complex and prolonged processing, puzzle feeders, object or sensory enrichment, and social access to conspecifics and other species may encourage the expression of more natural behavior patterns indoors.

THERMOREGULATION

Like elephants, giraffes are tropical animals, and temperatures in their natural habitat are relatively stable throughout the year. Because their long necks and legs give them a larger surface-to-volume ratio than the other tropical megaherbivores, giraffes are the most efficient of these species at dissipating heat (Mitchell & Skinner, 2004). However, in cool temperatures, giraffes' ability to lose heat puts them at a disadvantage, as even in cold weather they cannot increase insulation by building up fat reserves or increasing coat thickness (Langman, Langman, & Ellifrit, 2015). While they orient ". . . their bodies to optimize radiant heat gain and to maximize convective heat loss" (Mitchell & Skinner, 2004), Langman et al. (2015) found that elephants, okapi, and giraffes begin to draw on body reserves to maintain their core temperatures when the ambient temperature drops below 21°C (70°F). As Hummel, Zimmerman et al.'s (2006) survey reported that the middle 50% of European zoos maintained barn temperatures between 15–18°C during winter, it appears that zoos may be maintaining barn temperatures cold enough to cause giraffes significant thermal stress.

Despite being able to lose heat well, in hot temperatures, giraffes remain obligate heterotherms whose body temperature averages 38.5 ± 0.5°C but can vary by up to 5°C over the course of a day (Langman, Bamford, & Maloiy, 1982; Langman & Maloiy, 1989; Mitchell & Skinner, 2003, 2004). Using nasal heat loss, giraffes can keep their brain temperature significantly cooler than their body, reducing the neurological damage associated with heterothermy in other species (Mitchell &

Skinner, 2004). Outdoors at night, giraffes dissipate heat to the night sky to reestablish a neutral heat balance by morning (Langman & Maloiy, 1989; Mitchell & Skinner, 2003). Giraffe spots facilitate heat gain and loss by serving as "thermal windows;" sweat glands are packed densely in the spots, and surface blood vessels can constrict at their edges to regulate the amount of blood flow and therefore heat gain/loss (Mitchell & Skinner, 2004). When a giraffe's body temperature rises over 40°C (104°F), it begins to thermoregulate behaviorally by seeking shade (Langman & Maloiy, 1989). Smaller giraffes have more labile temperatures than do larger giraffes and spend more time performing thermoregulatory behaviors (Langman & Maloiy, 1989).

Captive environments can facilitate giraffe thermoregulation in a number of ways. In cool weather (less than 21°C), giraffes should be given ways to warm themselves behaviorally. Possible ideas for warming include having access to a heated barn, areas of the enclosure with heat-absorbing (dark-colored, gunite, stone) or heat-reflective surfaces (sand, light-colored, rock, concrete), exposure to direct sunlight, and shelter from wind and rain that can reduce body temperature further (Finegan, Atkinson, Buchanan-Smith, Cant, & Gillespie, 2003). Barn temperatures should be kept at a minimum of 21°C. In addition, the increased metabolic demands of thermoregulation during the winter may require feeding a higher-energy diet to reduce the risk of hypoglycemic collapse and death from "peracute mortality syndrome" (Potter & Clauss, 2005).

In hot weather (over 40°C), it is critical for captive giraffes to have as much outdoor access at night as possible to allow their body temperatures to decrease, as carrying a positive heat balance has been shown to alter metabolism, increase physiological stress, and decrease reproductive function in ruminants (Silanikove, 2000). In addition, giraffes should have access to cooler areas during the day so they can mitigate increases in body temperature. Such areas might include shade from trees or artificial structures, misters or air circulation, or even access to an air-conditioned space (Langman et al., 2003). In creating cool areas, care should be taken not to use heat-absorbing or -reflecting ground or wall substrates, which can provide enough radiant heat to reduce or negate the cooling effects of shade structures (Langman et al., 2003). Because most captive facilities use the same outdoor enclosure for their animals year-round, the ideal enclosure would provide a thermal mosaic (Gibson, Smucny, & Kollar, 1989), allowing giraffes to move among warm and cool microclimates. If undertaking this approach, ensure that giraffes have opportunities to eat in all microclimates (Arnold & Dudzinski, 1978) and have sufficient space in each area such that dominant giraffes cannot exclude subordinates.

FEEDING

Wild giraffes spend more than a quarter of their daylight hours feeding, though the exact amount of time varies by sex and location, e.g., 58% in Kruger National Park (du Toit & Yetman, 2005); males 27% and females 53% in Tsavo National Park (Leuthold & Leuthold, 1978); males 55% and females 72% in the Seronera woodlands (Pellew, 1984); and dry season 46% and rainy season 23% in Niger (Ciofolo & Le Pendu, 2002). In addition, Pellew (1984) reports giraffes in the Seronera woodlands fed for 22–34% of nighttime hours, with more feeding observed on brighter nights. Giraffes are browsers, primarily feeding on trees and shrubs, and the plant species that make up the vast majority of their diet are Acacia and Grewia. These species have both permanent defenses, such as large thorns and comparatively small leaves, and defenses activated by browsing, including tannin deployment and symbiotic stinging ants (Madden & Young, 1992). As a giraffe feeds on a plant with an activated defense system, that plant becomes progressively less appetizing (high tannin concentrations taste bad, and being stung by acacia ants is not a pleasant experience). Consequently, each giraffe tends to move from plant to plant while browsing rather than staying at a single plant until it is depleted, which improves the plants' rates of browse production (Pellew, 1983).

The last large-scale efforts to gather information about captive giraffe diets were in the early 2000s. Accredited American zoos were surveyed in 2000 (Bashaw et al., 2001), and European Zoos were surveyed in 2004 (Hummel, Zimmerman et al., 2006). Most zoos reported feeding their giraffes commercially manufactured diets (hereafter concentrates) either designed for ruminants (e.g., cattle) or specifically manufactured for browsers, supplemented with a combination of hays, fresh grass, and tree/shrub branches typically including leaves (hereafter browse). Some zoos in both regions also included fruits, vegetables, bread, or sweeter grain as part of the diet. Of these foods, browse most closely resembles the diet of wild giraffes, but it is not presented in sufficient quantity to be the primary food source for zoo-housed giraffe. In addition, the plant species used for browse are rarely as well-defended as the natural diet and have different nutritional composition. This feeding regime is still widely used and has two main problems: it may not be nutritionally appropriate, and it is associated with stereotypic behavior patterns that indicate compromised welfare.

Captive giraffe diets can lead to a variety of common health problems. Overconsumption of concentrates leaves giraffes vulnerable to rumen acidosis (Clauss, Kienzle, & Hatt, 2003) and frothy bloat (Colvile, Bouts, Hartley, Clauss, & Routh, 2009); overconsumption of grass, hay, or alfalfa provides insufficient energy for giraffes (Hatt et al., 2005) and may contribute to the development of bezoars (fibrous balls that can block the gastrointestinal tract: Clauss, Lechner-Doll, Flach, Wisser, & Hatt, 2002; Davis et al., 2009); and the consumption of "a higher proportion of easy digestible feeds (breads, pure grains, fruits, and vegetables)" predicts the occurrence of laminitis in giraffes (Hummel, Zimmerman et al., 2006). A healthy captive giraffe diet therefore requires limiting access to highly palatable concentrates, providing sufficient access to browse, and (if necessary) supplementing with additional roughage (alfalfa, grass, or hay) to maintain a minimum 60:40 ratio of roughage to concentrate (dry matter basis: Lintzenich & Ward, 1997). Zoos typically provide ad-lib access to hay or alfalfa, but giraffes are often reluctant to consume enough to achieve this ratio (Colvile et al., 2009; Hummel, Clauss, Baxter, Flach, & Johansen, 2006) because neither their teeth (Clauss, Franz-Odendaal, Brasch, Castell, & Kaiser, 2007) nor their digestive systems (Clauss et al., 2003) have the adaptations grazers use to digest grasses. Relying on concentrates to meet animals' nutritional needs can also be risky. Several Auckland Zoo giraffes died from underfeeding when energy and protein levels in their concentrated feed fell below the manufacturer's nutritional specifications during the winter (Potter & Clauss, 2005). Even when manufacturer specifications can be trusted, giraffes' nutritional needs are not well understood, so the ideal concentrate composition is yet unknown (Clauss et al., 2003). Several zoos have partnered with feed manufacturers in recent years to develop better concentrates for browsers, but work in this area is still in progress. Providing regular access to sufficient browse is critical for solving nutritional problems; even supplementing typical diets with browse brings fiber, protein, and carbohydrate consumption to more natural levels (Clauss et al., 2003; Hatt et al., 2005). "A year-round supply of browse would obviously be the desired feeding regime" (Clauss et al., 2002), but when this is not possible, browse should make up 10–25% of captive giraffe diets (Fowler, 1978). Although browse is expensive and in most locations its availability varies seasonally, regular access to at least some browse appears to be a nutritional requirement for giraffes (Clauss et al., 2003; Colvile et al., 2009). Hummel and Clauss (2006) have recommended zoos develop browse farms to ensure browse is readily available; browse exchange or freezing programs could also be used to increase giraffes' access to browse.

High rates of oral stereotypic behaviors (sensu: Mason, 2006) are observed in captive giraffes. Most giraffes perform oral stereotypic behaviors: 72.4% of captive giraffes and okapi in AZA-accredited zoos (Bashaw et al., 2001) and 160 of 161 individuals observed in Japanese zoos (Koene & Visser, 1997). Oral stereotypic behaviors common in giraffe (repetitive licking of nonfood objects and tongue playing: repeatedly rolling the tongue around inside and outside the mouth) resemble stereotypic behaviors associated with food-searching behavior in cattle (e.g., Redbo & Norblad, 1997). While tongue playing has been observed following eating and drinking in wild giraffes, it occupies an average of 1% of daylight observations in the wild and 11% (Bashaw, 2011) to 23% (Koene & Visser, 1997) of daylight observations in captivity. Oral stereotypic behaviors are more prevalent overnight, with giraffes at three zoos reported spending from 20% to more than 30% of overnight observations performing these behaviors (Duggan et al., 2016; Tarou et al., 2001). These studies also show giraffes housed in the same environment have very different rates of oral stereotypic behaviors.

Rates of oral stereotypic behaviors are affected by captive feeding regimes. The odds that a giraffe or okapi performed any oral stereotypic behaviors were significantly lower if the animal experienced less frequent feeding, was fed by staff only, was fed browse, was fed using closed-top feeders, had less night access to other giraffes (perhaps because of feeding competition), and spent fewer hours indoors (Bashaw et al., 2001). Experimental research has confirmed that feeding more browse (Koene & Visser, 1997), feeding more fiber (Baxter & Plowman, 2001), and increasing the oral effort required to obtain food (Fernandez, Bashaw, Sartor, Bouwens, & Maki, 2008) all reduce rates of oral stereotypic behaviors in giraffes. Several hypotheses have been posited to explain the relationship between feeding, rumination, and oral stereotypic behaviors in captive ungulates (for review, see Bergeron, Badnell-Waters, Lambton, & Mason, 2006). Some suggest oral stereotypic behaviors help giraffes physiologically cope with their captive diet by increasing saliva production, either to buffer acid in the rumen or to ensure adequate salivary phosphorus recycling (Ball, 2010). Others posit that oral stereotypic behaviors indicate giraffes are still hungry (or seeking a specific nutrient) after eating their captive diet. Still others argue that oral stereotypic behaviors are vacuum activities that fill the "spare time" captive animals gain by having easy access to food and water. Captive giraffes have substantial "spare time" because eating in captivity occupies only half as much time as it does in the wild (Baxter & Plowman, 2001; Koene, 1999). All of these probably contribute to the high rates of oral stereotypic behaviors in captive giraffes, which may explain why changing either nutrition or feeding technique can reduce them.

The three hypotheses about the function of oral stereotypic behaviors offer ideas for how to reduce these behaviors in captivity. Better matching giraffes' captive diet with their nutritional and behavioral needs (by providing more browse, more fiber, and better-formulated concentrates so we can reduce our reliance on hay or alfalfa) should ameliorate oral stereotypic behaviors to the extent that they are helping giraffes cope with poor captive diets or are seeking more food. Using feeders that slow the rate at which food is consumed should reduce oral stereotypic behaviors by both reducing "spare time" and decreasing spikes in rumen fermentation that are associated with discrete, rapidly consumed meals, particularly of concentrates (Hummel, Clauss et al., 2006). Giraffes' feeding time can be extended by making food more difficult to obtain (capping open-topped feeders, using puzzle feeders, hanging feeders nearly out of reach, or using feeders with openings small enough to exclude a giraffe's muzzle, as in Figure 1). Alternatively, zoos could design feeding regimes that require giraffe to move between patches as they would in the wild, for example by providing an array of feeders around the habitat that each have only a small quantity of food or building feeders that can move out of reach after a specific time or amount of food has been obtained. Combining an appropriate diet with feeders that slow consumption is likely to provide the most natural behavioral repertoire and healthiest giraffes.

FIG. 1 NATURALISTIC GIRAFFE FEEDER REQUIRING TONGUE MANIPULATION, CHICAGO ZOOLOGICAL SOCIETY/BROOKFIELD ZOO. PHOTO CREDIT: JIM SCHULTZ.

Public feeding programs (sometimes called guest feeding programs) have become a very popular part of giraffe exhibits over the past 20 years. On the face, these programs appear to benefit both zoo guests and giraffes. Guests benefit from the up-close animal encounter, which has been shown to increase positive emotion and engagement in zoo visitors (Powell & Bullock, 2014). As each guest typically provides only a small food item, giraffes have an opportunity to feed more slowly on a more dynamic food source. Our 2000 survey of AZA-accredited zoos found giraffes in zoos with public feeding had marginally lower odds of oral stereotypic behaviors, suggesting guest-feeding programs might benefit giraffes (Bashaw et al., 2001). An initial experimental study of the effects of guest-feeding programs on giraffe behavior in nine zoos in Australia found no positive or negative consequences on their welfare; the only significant behavioral change associated with public feeding was an increase in the time giraffes spent standing idle (Orban, Siegford, & Snider,

2016). However, the effects of guest feeding on giraffe behavior and welfare likely depend on the details of the program, particularly what is being fed to the giraffe and how. Conclusions in the feeding section above suggest guests should feed browse or other leafy greens rather than concentrates or fruit. Unless it is strictly roughage, what is fed should be considered part of the giraffes' diet and included in their nutritional plan. This may require adjusting the food the caretakers provide in response to attendance at public feeding stations to ensure that giraffes' nutritional requirements are met. Finally, no giraffe should be forced to interact with guests, so an additional source of equal quality food should be available to allow each giraffe to choose whether or not to participate. Further study of guest-feeding programs is ongoing and may result in changes to these recommendations.

SOCIAL RELATIONSHIPS

Both African elephants and giraffes have fission-fusion social structures. Female African elephants live in fairly stable family units (core groups) composed of maternally related individuals and their juvenile offspring. These units form the second tier of a multilevel social structure within which ecological constraints affect the combination and separation of groups and, to a lesser extent, individuals (Archie, Moss, & Alberts, 2006; Wittemyer, Douglas-Hamilton, & Getz, 2005). Female giraffe social structure is similar to African elephants in that group composition varies within a multilevel social structure (VanderWaal, Wang, McCowan, Fushing, & Isbell, 2014) structured around genetic relatedness (particularly mother/daughter bonds), affiliation with older females, and peer relationships (Bercovitch & Berry, 2013a, 2013b; Carter, Seddon, Frère, Carter, & Goldizen, 2013). However, compared to African elephants, giraffe groups are less cohesive, group membership is less consistent, and multigenerational associations are rarer (Bercovitch & Berry, 2013a, 2013b; Shorrocks & Croft, 2009). Despite looser dyadic relationships, individual giraffe females actively maintain relationships with one another, and association patterns cannot simply be explained by spatial overlap (Malyjurkova, Hejzlarova, Vymyslicka, & Brandlova, 2014; VanderWaal et al., 2014). Male life histories also show similarities and differences. Both giraffe and elephant males are found in a variety of social groups before maturity, but by full maturity, males travel alone, using a roving strategy to locate fertile females. Giraffe males compete for females both by forming a dominance hierarchy and undergoing short rut-like periods of elevated testosterone (Seeber, Duncan, Fritz, & Ganswindt, 2013), though these periods are much less distinct, shorter in duration, and less regular than male elephants' musth (Poole & Moss, 1981).

Captive giraffe herds housed in large exhibits have similar social interaction patterns to those observed in the wild, with individuals frequently changing social partners but exhibiting consistent preferences over time (Bashaw et al., 2007). The difference between male and female social structure in the wild suggests that males and females in captivity should be managed differently. For females, the most natural social grouping would be to keep herds of six to eight individuals based on matrilines and to split these herds along maternal lines when they grow too large for a single enclosure. Males should be housed in larger enclosures, encouraged to travel, and potentially even allowed to seek access to females or their scents. The presence of dominance hierarchies among male (Coe, 1967; Leuthold, 1979; Pratt & Anderson, 1985) and captive female giraffes (Horová, Brandlová, & Gloneková, 2015) suggests subordinate animals will value opportunities to retreat from social interactions and sufficient sources of important resources such that dominant animals cannot monopolize them (as described for bachelor gorillas in Coe, Scott, & Lukas, 2009). It is possible that changing what and how we feed giraffes might also improve subordinates' access to food; Plowman and Cabana (in press) found that when sugary fruits were removed from the diets of Sulawesi crested black macaques and Diana monkeys in favor of a more naturalistic diet, dominant animals no longer monopolized food resources. Ideally, enclosures containing groups of giraffes should have sufficient space and visual barriers to allow individuals to regulate social contact within the herd. Although captive groups provided with abundant food and water sources may elect to stay together most of the time, the opportunity to escape from or avoid conspecifics is likely to improve welfare even if it is rarely used (Kurtycz, Wagner, & Ross, 2014).

APPLYING PREDICTABILITY AND CONTROL TO IMPROVE WELLNESS

Environmental Events and Schedules

Having a consistent, reliable cue for an event allows an animal to anticipate when that event will occur and prepare for it. Bassett and Buchanan-Smith (2007) provided an excellent review of the effects of predictability on welfare, so I will give only a brief summary here. For aversive or startling events, a reliable cue that precedes the events allows the animal to mobilize behavioral and physiological coping strategies before they are needed. Animals regularly transfer physiological compensatory responses (activation of the stress response system, release of endogenous opioids, etc.) from negative events to the cues that predict them so that after a few experiences escaping predictable aversive events, animals learn to avoid them (Powell, Honey, & Symbaluk, 2016). Because this system is quite reliable, providing a discrete noticeable cue (for example, a whistle, colored light, or soft music) before each instance of an event you expect may be negative (perhaps a loud concert or fireworks show) will allow the animals to escape and eventually avoid the negative effects of the event by deploying behavioral and physiological coping mechanisms. For example, Rimpley and Buchanan-Smith (2013) found that knocking before entering brown capuchin enclosures reduced anxiety-related behaviors and increased the capuchin's welfare. Disney Animal Kingdom's use of a soft sound at the beginning of events (described above in "Sound") is an excellent example of this strategy. Giraffes may particularly benefit from the opportunity to anticipate events with unusual amounts of noise (like a concert) or a substantial visual component (like a parade).

For desirable (appetitive) events, the effects of predictability are less clear-cut. Appetitive predictability has mostly been studied in relation to feeding. Anticipating a feeding just before it arrives makes that feeding more pleasurable, but anticipating a feeding that does not arrive is aversive. As a result, providing a reliable cue (for example, consistently blowing a whistle immediately before feeding, as at Elmwood Park Zoo) improves welfare, while having only unreliable cues (e.g., a caretaker walking up to the exhibit or entering the barn, food preparation noise) to predict feeding can compromise welfare. Note that time cues (e.g., feeding at the same time each day) are inherently unreliable because animals cannot tell time very precisely. Unreliable cues lead to poor welfare by generating persistent unsatisfied food anticipation and can also trigger pacing or oral stereotypic behavior that is eventually reinforced by feeding. In contrast, feeding at unpredictable times and/or in unpredictable places reinforces exploration and investigation, which reflect improved welfare. If feeding must occur at about the same time every day because of keeper schedules, adding a more reliable cue immediately before feeding can limit anticipation. Devices that increase consumption time or conceal food delivery or availability could also be used to make feeding a less discrete and exciting event. Efforts to increase predictability are especially important for highly palatable foods like concentrates.

Choice and control have long been recognized as crucial to welfare but have not typically been differentiated. Allard and Bashaw (in press) review these concepts and their applications to animal care in zoos. In short, providing animals with variety within their enclosure and routine increases choice by creating a variety of behavioral options that animals can decide how to use. Several of the suggestions made earlier in this chapter allow giraffes to choose various aspects of their own lives without the need for interference by humans. For example, providing an exhibit with a variety of microclimates allows an animal choice in thermoregulation, and providing a variety of types of feed allows some choice in feeding. Since animals choose to choose and will work to obtain access to choice, we can infer they value having choices and should provide choices whenever possible.

Control allows an animal to cause, stop, or prevent an environmental event rather than simply change their behavior to prepare for or cope with the inevitable. Buchanan-Smith and Badihi (2012) found that having control over light and heat improved the welfare of captive marmosets more than just having access to light and heat for the same duration. Giving an animal the opportunity to work for a stimulus allows caretakers to measure how much an animal values that stimulus. For example, a giraffe with the opportunity to travel to locate a novel scent or social partner can demonstrate how desirable that scent or partner is by how far s/he was willing to go to obtain it. Finally, responsive management, in which caretakers attend to an animal's behavior and use cues from the animal to determine what areas of the habitat will be available, what type of food will be provided, or whether or not to interact with the animal, can allow that animal a degree of control over its own care. Sambrook and Buchanan-Smith (1997) suggested that the acquisition of control is more enriching than exercising control, so animal managers should search continually for new ways to allow animals to gain control or change the behavior required to achieve a particular goal to continue to challenge animals. Of course, too much predictability and control can result in boredom; thus, it is critical to strike a dynamic balance of predictability and unpredictability, and of having and not having control (Watters and Vicino, 2016).

INTERACTIONS WITH HUMANS

Interactions with humans are one context in which balancing animals' experiences of predictability and control is particularly important. Although giraffes do not interact regularly with humans in the wild, in zoos, they depend on human caretakers for many aspects of their lives and are exposed to zoo guests daily. For this reason, Hosey (2008) suggested that creating positive (or at least neutral) human-animal relationships with both caretakers and zoo guests is critical for the well-being of zoo animals.

Because the same caretakers have repeated interactions with their giraffes, they can build positive (or negative) relationships with them. Studies of domesticated ruminants have shown that stockmanship is an important determinant of animal well-being (e.g., Boivin, Lensink, Tallet, & Veissier, 2003; Hemsworth, 2007) and suggested assessment of human-animal relationships be integrated into welfare monitoring regulations (Rennie, Bowell, Dearing, Haskell, & Lawrence, 2003; Hemsworth, Barnett, & Coleman, 2009). Zoo-housed ungulates attend to the presence of caretakers (Thompson, 1989). Animal managers can harness this attention to improve wellness by facilitating predictable, positive interactions with caretakers (for review, see Claxton, 2011). Having a small, consistent caretaker staff gives zoo animals more experience with the behavior of each individual caretaker, rendering caretaker behavior more predictable for the animal, mitigating animals' innate fear of humans (Hosey, 2008) and potentially improving well-being. For example, Koester (2014) found that cheetahs cared for by three or fewer caretakers were healthier. If a small caretaker staff is not possible, assigning each caretaker to interact with a particular individual giraffe on a regular basis could achieve some of the same effects. Animals also can be given some control over their relationships with caretakers using positive reinforcement training across a barrier (i.e., protected contact management: Laule, 2003). In this system, animals are reinforced for completing desired behaviors according to a set of rules but are free to terminate the session or choose not to participate at any time with no negative consequences. Protected contact interactions can not only improve psychological well-being but also physical well-being. For example, staff at Cheyenne Mountain Zoo used this training scheme to obtain radiographs of their giraffes' hooves and improve diagnosis and treatment of medical issues (Dadone et al., 2016).

Giraffes have limited opportunities for repeated interactions with the same zoo guest, so guests (unlike caretakers) are likely to be perceived as members of a category rather than as individuals (Hosey, 2008).

Guests may be perceived as a threat by some smaller ungulates, but for giraffes, their most aversive quality may be noisiness. Positive feelings about guests could be enhanced by using architectural techniques to encourage guests to remain quiet and by creating more opportunities for guests to have positive interactions with giraffes (e.g., using guest-feeding programs). Giraffes' perceptions of guests could also be improved by placing interactions with guests more under the giraffe's control (for example, by ensuring equally desirable food is available elsewhere in the exhibit during public feeding) and allowing giraffes to avoid negative interactions (e.g., by providing retreat space: Anderson, Benne, Bloomsmith, & Maple, 2002).

CONCLUSIONS

In designing animal housing spaces and management programs for optimal wellness, it is important to consider the animal's point of view. This includes considering what they can perceive and what they are motivated to do in addition to their physiological needs. For giraffes, access to visual and olfactory information, the opportunity to avoid noisy environments, and adequate opportunities to sleep, thermoregulate, and socialize are particularly important considerations for wellness. Feeding practices for giraffes have improved over the past two decades, but persistent reports of oral stereotypic behavior and digestive maladies suggest continued room for improvement. Similarly, more research is required to optimize housing and management protocols, particularly overnight and during the winter months. When implemented carefully, positive reinforcement training and guest-feeding programs show promise as ways to allow giraffes to form more positive relationships with humans.

REFERENCES

Aidala, Z., Huynen, L., Brennan, P. L., Musser, J., Fidler, A., Chong, N., & Hauber, M. E. (2012). Ultraviolet visual sensitivity in three avian lineages: Paleognaths, parrots, and passerines. *Journal of Comparative Physiology A, 198,* 495–510.

Allard, S. M., & Bashaw, M. J. (2019) Empowering zoo animals. In A. B. Kaufman, M. J. Bashaw, & T. L. Maple (Eds.), *Scientific Foundations of Zoos and Aquariums: Their role in conservation and research (pp. 241-273).* Cambridge University Press.

Anderson, U. S., Benne, M., Bloomsmith, M. A., & Maple, T. L. (2002). Retreat space and human guest density moderate undesirable behavior in petting zoo animals. *Journal of Applied Animal Welfare Science, 5*(2), 125–137.

Archie, E. A., Moss, C. J., & Alberts, S. C. (2006). The ties that bind: Genetic relatedness predicts the fission and fusion of social groups in wild African elephants. In *Proceedings of the Royal Society of London B: Biological Sciences, 273,* 513–522.

Arnold, G. W., & Dudzinski, M. L. (1978). *Ethology of free-ranging domestic animals.* New York, NY: Elsevier Scientific.

Ball, R. L. (2010). Clinical issues associated with nutrition and feeding in managed giraffe. *Giraffa, 4*(1), 50.

Baotic, A., Sicks, F., & Stoeger, A. S. (2015). Nocturnal "humming" vocalizations: Adding a piece to the puzzle of giraffe vocal communication. *BMC Research Notes, 8,* 1.

Bargai, U., and Cohen, R. (1992). Tarsal lameness of dairy bulls housed at two artificial insemination centers: 24 cases (1975–1987). *Journal of the American Veterinary Medical Association, 201,* 1,068–1,069.

Bashaw, M. J. (2003). *Social behavior and communication in a herd of captive giraffe* (Doctoral dissertation). Retrieved from https://smartech.gatech.edu/bitstream/handle/1853/5386/bashaw_meredith_j_200312_phd.pdf.

Bashaw, M. J. (2011). Consistency of captive giraffe behavior under two different management regimes. *Zoo Biology, 30,* 371–378.

Bashaw, M. J., Bloomsmith, M. A., Maple, T. L., & Bercovitch, F. B. (2007). The structure of social relationships among captive female giraffe. *Journal of Comparative Psychology, 121,* 46–53.

Bashaw, M. J., Tarou, L. R., Maki, T. S., & Maple, T. L. (2001). A survey assessment of variables related to stereotypy in giraffe and okapi. *Applied Animal Behavior Science, 73,* 233–245.

Bassett, L., & Buchanan-Smith, H. M. (2007). Effects of predictability on the welfare of captive animals. *Applied Animal Behavior Science, 102,* 223–245.

Baxter, E., & Plowman, A. B. (2001). The effect of increasing dietary fiber on feeding, rumination, and oral stereotypies in captive giraffes (*Giraffa camelopardalis*). *Animal Welfare,* 10, 281–290.

Bercovitch, F. B., Bashaw, M. J., & del Castillo, S. M. (2006). Sociosexual behavior, male mating tactics, and the reproductive cycle of giraffe, *Giraffa camelopardalis. Hormones & Behavior, 50,* 314–321. doi: 10.1016/j.yhbeh.2006.04.004

Bercovitch, F. B., & Berry, P. S. M. (2013a). Age proximity influences herd composition in wild giraffe. *Journal of Zoology, 290*, 281–286.

Bercovitch, F. B., & Berry, P. S. (2013b) Herd composition, kinship, and fission-fusion social dynamics among wild giraffe. *African Journal of Ecology, 51*, 206–216.

Bercovitch, F. B. & Deacon, F. (2015). Gazing at a giraffe gyroscope: Where are we going? *African Journal of Ecology, 53*, 135–146. doi: 10.1111/aje.12222

Bergeron, R., Badnell-Waters, A. J., Lambton, S., & Mason, G. (2006). Stereotypic oral behavior in captive ungulates: Foraging, diet, and gastrointestinal function. In G. Mason & J. Rushen (Eds.), *Stereotypic animal behavior: Fundamentals and applications to welfare* (pp. 19–41). Cambridge, MA: CAB International.

Boivin, X., Lensink, J., Tallet, C., & Veissier, I. (2003). Stockmanship and farm animal welfare. *Animal Welfare, 12*, 479–492.

Branch, M. C., & Beland, R. D. (1970). *Outdoor noise and the metropolitan environment: Case study of Los Angeles, with special reference to aircraft.* Los Angeles Department of City Planning.

Buchanan-Smith, H. M., & Badihi, I. (2012). The psychology of control: Effects of control over supplementary light on welfare of marmosets. *Applied Animal Behavior Science, 137*, 166–174.

Cameron, E. Z., & du Toit, J. T. (2005). Social influences on vigilance behavior in giraffes, *Giraffa camelopardalis. Animal Behavior, 69*, 1,337–1,344.

Carter, K. D., Seddon, J. M., Frère, C. H., Carter, J. K., & Goldizen, A. W. (2013). Fission-fusion dynamics in wild giraffes may be driven by kinship, spatial overlap, and individual social preferences. *Animal Behavior, 85*, 385–394.

Ciofolo, I., & Le Pendu, Y. (2002). The feeding behavior of giraffe in Niger. *Mammalia, 66*, 183–194.

Clauss, M., Franz-Odendaal, T. A., Brasch, J., Castell, J. C., & Kaiser, T. (2007). Tooth wear in captive giraffes (*Giraffa camelopardalis*): Mesowear analysis classifies free-ranging specimens as browsers but captive ones as grazers. *Journal of Zoo and Wildlife Medicine, 38*, 433–445.

Clauss, M., Kienzle, E., & Hatt, J. M. (2003). Feeding practice in captive wild ruminants: Peculiarities in the nutrition of browsers/concentrate selectors and intermediate feeders. A review. *Zoo Animal Nutrition, 2*, 27–52.

Clauss, M., Lechner-Doll, M., Flach, E. J., Wisser, J., & Hatt, J.-M. (2002). Digestive tract pathology of the giraffe (*Giraffa camelopardalis*). An unifying hypothesis. *Proceedings of the 4th Scientific Meeting of the European Association of Zoo and Wildlife Veterinarians* (pp. 99–107). Heidelberg, Germany: EAZWV.

Claxton, A. M. (2011). The potential of the human-animal relationship as an environmental enrichment for the welfare of zoo-housed animals. *Applied Animal Behavior Science, 133*, 1–10.

Coe, M. J. (1967). "Necking" behavior in the giraffe. *Journal of Zoology, 151*(3), 313–321.

Coe, J. C., Scott, D., & Lukas, K. E. (2009). Facility design for bachelor gorilla groups. *Zoo Biology, 28*(2), 144–162.

Coimbra, J. P., Hart, N. J., Collin, S. P., & Manger, P. R. (2013). Scene from above: Retinal ganglion cell topography and spatial resolving power in the giraffe (*Giraffa camelopardalis*). *Journal of Comparative Neurology, 521*, 2,042–2,057.

Colvile, K., Bouts, T., Hartley, A., Clauss, M., & Routh, A. (2009). Frothy bloat and serous fat atrophy in a giraffe (*Giraffa camelopardalis*) with chronic respiratory disease. In M. Clauss, A. Fidgett, J. M. Hatt, T. Huisman, J. Hummel, G. Janssen, J. Nijboer, & A. Plowman, *Zoo animal nutrition* (Vol. IV) (pp. 219–229). Fürth, Germany: Filander.

Dadone, L. I., Schilz, A., Friedman, S. G., Bredahl, J., Foxworth, S., & Chastain, B. (2016). Training giraffe (*Giraffa camelopardalis reticulata*) for front foot radiographs and hoof care. *Zoo Biology, 35*, 228–236.

Dagg, A. I. (2014). *Giraffe: Biology, behavior, and conservation*. United Kingdom: Cambridge University Press.

Davis, M. R., Langan, J. N., Mylniczenko, N. D., Benson, K., Lamberski, N., & Ramer, J. (2009). Colonic obstruction in three captive reticulated giraffe (*Giraffa camelopardalis reticulata*). *Journal of Zoo and Wildlife Medicine, 40*, 181–188.

Duggan, G., Burn, C. C., & Clauss, M. (2016). Nocturnal behavior in captive giraffe (*Giraffa camelopardalis*)—A pilot study. *Zoo Biology, 35*, 14–18. doi: 10.1002/zoo.21248

Du Toit, J. T., & Yetman, C. A. (2005). Effects of body size on the diurnal activity budgets of African browsing ruminants. *Oecologia, 143*, 317–325.

Fay, C., & Miller, L. J. (2015). Utilizing scents as environmental enrichment: Preference assessment and application with Rothschild giraffe. *Animal Behavior and Cognition, 2*, 285–291.

Fernandez, L. T., Bashaw, M. J., Sartor, R. L., Bouwens, N. R., & Maki, T. S. (2008). Tongue twisters: Feeding enrichment to reduce oral stereotypy in giraffe. *Zoo Biology, 27*, 200–212. doi: 10.1002/zoo.20180

Finegan, E. J., Atkinson, J. L., Buchanan-Smith, J. G., Cant, J. P., & Gillespie, T. J. (2003.) Thermal constraints on grazing and browsing herbivores. Nutrition Advisory Group. Retrieved from https://www.researchgate.net/profile/John_Cant/publication/237293257_THERMAL_CONSTRAINTS_ON_GRAZING_AND_BROWSING_HERBIVORES/links/542157bc0cf2ce3a91b70b76.pdf.

Fowler, M. E. (1978). Peracute mortality in captive giraffe. *Journal of the American Veterinary Medical Association, 173*, 1,088–1,093.

Fraser, D. (2009). Assessing animal welfare: Different philosophies, different scientific approaches. *Zoo Biology, 28*, 507–518.

Gelez, H., & Fabre-Nys, C. (2004). The "male effect" in sheep and goats: A review of the respective roles of the two olfactory systems. *Hormones and Behavior, 46*, 257–271.

Gibson, A. R., Smucny, D. A., & Kollar, J. (1989). The effects of feeding and ecdysis on temperature selection by young garter snakes in a simple thermal mosaic. *Canadian Journal of Zoology, 67*, 19–23.

Hatt, J.-M., Schaub, D., Wanner, M., Wettstein, H.-R., Flach, E. J., Tack, C., . . . Clauss, M. (2005). Energy and fibre intake in a group of captive giraffe (*Giraffa camelopardalis*) offered increasing amounts of browse. *Journal of Veterinary Medicine Series A, 52,* 485–490. doi: 10.1111/j.1439-0442.2005.00769.x

Hemsworth, P. H. (2007). Ethical stockmanship. *Australian Veterinary Journal, 85*(5), 194–200.

Hemsworth, P. H., Barnett, J. L., & Coleman, G. J. (2009). The integration of human-animal relations into animal welfare monitoring schemes. *Animal Welfare, 18,* 335–345.

Horová, E., Brandlová, K., & Gloneková, M. (2015). The first description of dominance hierarchy in captive giraffe: Not loose and egalitarian, but clear and linear. *PLOS ONE, 10*(5), e0124570.

Hosey, G. (2008). A preliminary model of human-animal relationships in the zoo. *Applied Animal Behavior Science, 109,* 105–127.

Hummel, J., & Clauss, M. (2006). Feeding. In EAZA Giraffe EEPs (Eds.), *EAZA husbandry and management guidelines for* Giraffa camelopardalis (pp. 29–61). Arnhem, the Netherlands: Burger's Zoo.

Hummel, J., Clauss, M., Baxter, E., Flach, E. J., & Johansen, K. (2006). The influence of roughage intake on the occurrence of oral disturbances in captive giraffids. *Zoo Animal Nutrition, 3,* 235–252.

Hummel, J., Zimmerman, W., Langenhorst, T., Schleussner, G., Damen, M., & Clauss, M. (2006). Giraffe husbandry and feeding practices in Europe: Results of an EEP survey. *Proceedings of the European Association of Zoo and Wildlife Veterinarians 6th Scientific Meeting*, Budapest, Hungary.

Koene, P. (1999). When feeding is just eating: How do farm and zoo animals use their spare time? In D. van der Heide, E. A. Huisman, E. Kanis, J. W. M. Osse, & M. W. A. Verstegen (Eds.), *Regulation of feed intake: Proceedings of the 5th Zodiac Symposium* (pp. 13–20). Wageningen, the Netherlands: CAB.

Koene P., & Visser, E. K. (1997). Tongue playing behavior in captive giraffes [Supplementum II]. *Zeitschrift fur Saugetierkunde, 62,* 106–111.

Koester, D. C. (2014). *The effect of environment on the reproductive potential of ex xitu cheetahs (Acinonyx jubatus)* (Doctoral dissertation). George Mason University, Fairfax, VA.

Kurtycz, L. M., Wagner, K. E., and Ross, S. R. (2014). The choice to access outdoor areas affects the behavior of great apes. *Journal of Applied Animal Welfare Science, 17,* 185–197.

Langman, V. A. (1977). Cow-calf relationships in giraffe (*Giraffa camelopardalis giraffa*). *Zeitschrift für Tierpsychologie, 43,* 264–286.

Langman, V. A., Bamford, O. S., & Maloiy, G. M. O. (1982). Respiration and metabolism in the giraffe. *Respiration Physiology, 50,* 141–152.

Langman, V. A., Langman, S. L., & Ellifrit, N. (2015). Seasonal acclimatization determined by non-invasive measurements of coat insulation. *Zoo Biology,* 34, 368–373.

Langman, V. A., & Maloiy, G. M. O. (1989). Passive obligatory heterothermy of the giraffe. *Journal of Physiology (UK), 415,* 89.

Langman, V. A., Rowe, M., Forthman, D., Langman, N., Black, J., & Walker, T. (2003). Quantifying shade using a standard environment. *Zoo Biology, 22,* 253–260.

Laule, G. E. (2003). Positive reinforcement training and environmental enrichment: Enhancing animal well-being. *Journal of the American Veterinary Medical Association, 223,* 969–973.

Leuthold, B. M. (1979). Social organization and behaviour of giraffe in Tsavo East National Park. *African Journal of Ecology, 17,* 19–34.

Leuthold, B. M., & Leuthold, W. (1978). Daytime activity patterns of gerenuk and giraffe in Tsavo National Park, Kenya. *African Journal of Ecology, 16,* 231–243.

Lintzenich, B. A., & Ward, A. M. (1997.) Hay and pellet ratios: Considerations in feeding ungulates. *NAG Handbook Fact Sheet 006.* Silver Spring, MD: Nutrition Advisory Group, American Zoo and Aquarium Association.

Madden, D., & Young, T. P. (1992). Symbiotic ants as an alternative defense against giraffe herbivory in spinescent *Acacia drepanolobium. Oecologia, 91,* 235–238.

Malyjurkova, L., Hejzlarova, M., Vymyslicka, P. J., & Brandlova, K. (2014). Social preferences of translocated giraffes (*Giraffa camelopardalis giraffa*) in Senegal: Evidence for friendship among females? *Agricultura Tropica et Subtropica, 47,* 5–13.

Mason, G. (2006). Stereotypic behaviour in captive animals: Fundamentals, and implications for welfare and beyond. In G. Mason & J. Rushen (Eds.), *Stereotypic behavior in captive animals: Fundamentals and applications for welfare* (2nd ed.) (pp. 325–356). Wallingford, United Kingdom: CAB.

Mitchell, G., Roberts, D. G., Van Sittert, S. J., & Skinner, J. D. (2013). Orbit orientation and eye morphometrics in giraffes (*Giraffa camelopardalis*). *African Zoology, 48*, 333–339.

Mitchell, G., & Skinner, J. D. (2003). On the origin, evolution, and phylogeny of giraffes (*Giraffa camelopardalis*). *Transactions of the Royal Society of South Africa, 58*, 51–73.

Mitchell, G., & Skinner, J. D. (2004). Giraffe thermoregulation: A review. *Transactions of the Royal Society of South Africa, 59*, 109–118.

Morgan, K. N., & Tromborg, C. T. (2007). Sources of stress in captivity. *Applied Animal Behavior Science, 102*(3), 262–302.

Nagel, T. (1974). What is it like to be a bat? *The Philosophical Review, 83*, 435–450.

Norring, M., Manninen, E., De Passille, A. M., Rushen, J., & Saloniemi, H. (2010). Preferences of dairy cows for three stall surface materials with small amounts of bedding. *Journal of Dairy Science, 93*, 70–74.

Orban, D. A., Siegford, J. M., & Snider, R. J. (2016). Effects of guest feeding programs on captive giraffe behavior. *Zoo Biology, 35*, 157–166.

Pellew, R. A. (1983). The giraffe and its food resource in the Serengeti. I. Composition, biomass, and production of available browse. *African Journal of Ecology, 21*, 241–267.

Pellew, R. A. (1984). The feeding ecology of a selective browser, the giraffe (*Giraffa camelopardalis tippelskirchi*). *Journal of Zoology, 202*, 57–81.

Plowman, A., & Cabana, F. (2019). Transforming the nutrition of zoo primates (or how we became known as Loris Man and that evil banana woman). In A. B. Kaufman, M. J. Bashaw, & T. L. Maple (Eds.), *Scientific Foundations of Zoos and Aquariums: Their role in conservation and research (pp. 274-303).* Cambridge, United Kingdom: Cambridge University Press.

Poole, J. H., & Moss, C. J. (1981). Musth in the African elephant, *Loxodonta africana. Nature, 292*, 830–831.

Potter, J. S., & Clauss, M. (2005). Mortality of captive giraffe (*Giraffa camelopardalis*) associated with serous fat atrophy: A review of five cases at Auckland Zoo. *Journal of Zoo and Wildlife Medicine, 36*, 301–307.

Powell, D. M., & Bullock, E. V. (2014). Evaluation of factors affecting emotional responses in zoo guests and the impact of emotion on conservation mindedness. *Anthrozoös, 27*, 389–405.

Powell, R. A., Honey, P. L., & Symbaluk, D. G. (2016). *Introduction to learning and behavior* (5th ed.). Boston, MA: Cengage Learning.

Pratt, D. M., & Anderson, V. H. (1985). Giraffe social behavior. *Journal of Natural History, 19*, 771–781.

Redbo, I., & Norblad, A. (1997). Stereotypies in heifers are affected by feeding regime. *Applied Animal Behavior Science, 53*, 193–202.

Rennie, L. J., Bowell, V. A., Dearing, J. M., Haskell, M. J., & Lawrence, A. B. (2003). A study of three methods used to assess stockmanship on commercial dairy farms: Can these become effective welfare assessment techniques? *Animal Welfare, 12*, 591–597.

Rimpley, K., & Buchanan-Smith, H. M. (2013). Reliably signalling a startling husbandry event improves welfare of zoo-housed capuchins (*Sapajus apella*). *Applied Animal Behavior Science, 147*, 205–213.

Ross, M. R., Gillespie, K. L., Hopper, L. M., Bloomsmith, M. A., & Maple, T. L. (2013). Differential preference for ultraviolet light among captive birds from three ecological habitats. *Applied Animal Behavior Science, 147*, 278–285.

Sambrook, T. D., & Buchanan-Smith, H. M. (1997). Control and complexity in novel object enrichment. *Animal Welfare, 6*, 207–216.

Seeber, P. A., Ciofolo, I., & Ganswindt, A. (2012). Behavioral inventory of the giraffe (*Giraffa camelopardalis*). *BMC Research Notes, 5*, 1.

Seeber, P. A., Duncan, P., Fritz, H., & Ganswindt, A. (2013). Androgen changes and flexible rutting behavior in male giraffes. *Biology Letters, 9*, 20130396. doi: 10.1098/rsbl.2013.0396

Shepherdson, D., Lewis, K. D., Carlstead, K., Bauman, J., & Perrin, N. (2013). Individual and environmental factors associated with stereotypic behavior and fecal glucocorticoid metabolite levels in zoo-housed polar bears. *Applied Animal Behavior Science, 147*, 268–277.

Shorrocks, B., & Croft, D. P. (2009) Necks and networks: A preliminary study of population structure in the reticulated giraffe (*Giraffa camelopardalis reticulata* de Winston). *African Journal of Ecology, 47*, 374–381.

Sicks, F. (2012). *Paradoxer schlaf als parameter zur messung der stressbelastung bei giraffen* (Giraffa camelopardalis) (Unpublished doctoral dissertation). Johann Wolfgang Goethe-Universität, Frankfurt am Main.

Sicks, F. (2016). REM sleep as indicator for stress in giraffes (*Giraffa camelopardalis*). *Mammalian Biology-Zeitschrift für Säugetierkunde, 81S*, 16.

Siegel, J. M. (2005). Clues to the functions of mammalian sleep. *Nature, 437*, 1,264–1,271.

Silanikove, N. (2000). Effects of heat stress on the welfare of extensively managed domestic ruminants. *Livestock Production Science, 67*, 1–18.

Tarou, L. R., Bashaw, M. J., Sartor, R., Maki, T. S., Liu, S., & Maple, T. L. (2001). *When we're not looking: Nocturnal behavior in giraffe.* Paper presented at the national meeting of the American Zoo and Aquarium Association, St. Louis, MO.

Telezhenko, E., & Bergsten, C. (2005). Influence of floor type on the locomotion of dairy cows. *Applied Animal Behavior Science, 93*, 183–197.

Thompson, V. D. (1989). Behavioral response of 12 ungulate species in captivity to the presence of humans. *Zoo Biology, 8*, 275–297.

Tobler, I. (1992). Behavioral sleep in the Asian elephant in captivity. *Sleep, 15*, 1–12.

Tobler, I., & Schwierin, B. (1996). Behavioral sleep in the giraffe (*Giraffa camelopardalis*) in a zoological garden. *Journal of Sleep Research, 5*, 21–32.

Tromborg, C. T., & Coss, R. C. (1995). Denizens, decibels and dens. *Annual Proceedings of the American Association of Zoos and Aquariums* (pp. 521–528). Seattle, WA: AZA.

Tucker, C. B., Weary, D. M., Von Keyserlingk, M. A. G., & Beauchemin, K. A. (2009). Cow comfort in tie-stalls: Increased depth of shavings or straw bedding increases lying time. *Journal of Dairy Science, 92*, 2,684–2,690.

Van der Tol, P. P. J., Metz, J. H. M., Noordhuizen-Stassen, E. N., Back, W., Braam, C. R., & Weijs, W. A. (2005). Frictional forces required for unrestrained locomotion in dairy cattle. *Journal of Dairy Science, 88*, 615–624.

VanderWaal, K. L., Wang, H., McCowan, B., Fushing, H., & Isbell, L. A. (2014). Multilevel social organization and space use in reticulated giraffe (*Giraffa camelopardalis*). *Behavioral Ecology, 25*, 17–26.

von Muggenthaler, E. (2013). Giraffe Helmholtz resonance. *Proceedings of Meetings on Acoustics, 19*, 010012. doi: 10.1121/1.4800658

von Uexküll, J. (1934). A stroll through the worlds of animals and men: A picture book of invisible worlds. In C. H. Schiller (Ed.), *Instinctive behavior: Development of a modern concept* (pp. 5–80). New York, NY: International Universities Press.

Waser, P. M., and Brown, C. H. (1986). Habitat acoustics and primate communication. *American Journal of Primatology, 10,* 135–154.

Watters, J., & Vicino, G. (2016, September). *Balancing routine and unpredictability to support welfare.* Paper presented at the annual conference of the Association of Zoos and Aquariums, San Diego, CA.

Wittemyer, G., Douglas-Hamilton, I., & Getz, W. M. (2005). The socioecology of elephants: Analysis of the processes creating multitiered social structures. *Animal Behavior, 69,* 1,357–1,371.

Wood, W. F., & Weldon, P. J. (2002). The scent of the reticulated giraffe (*Giraffa camelopardalis reticulata*). *Biochemical Systematics and Ecology, 30,* 913–917.

Chapter 16

REPTILES NEED ENRICHMENT TOO, AND NOT JUST MONITOR LIZARDS

A Behavior Systems Approach

Gordon M. Burghardt
Departments of Psychology and Ecology & Evolutionary Biology

INTRODUCTION

C aptive reptiles are largely ignored in terms of the types of cognitive and behavioral training and enrichment often provided to mammals and birds. Although naturalistic habitats are common in exhibit facilities, often these are developed from a human-centric, not reptile-centric, perspective. Many zoos are now providing more challenging enrichment opportunities, and this trend should continue. This chapter presents an approach to captive reptiles and other animals, embedded in a behavior systems model of behavioral organization.

As the lone person at the meeting at the Jacksonville Zoo and Gardens Elephant Wellness Workshop asked to discuss a nonmammalian group, let alone elephants, I was a bit perplexed as to my role. Then a few observations emerged. One is that many nonavian reptiles (reptiles from here on) kept at zoos are large and potentially dangerous, and thus, routines and procedures for handling them and improving the quality of their lives in captivity present some useful parallels and contrasts to maintaining successfully other difficult and potentially dangerous species, such as elephants. Second, the reptile-mammal transition, although going back to the therapsids in the Triassic Period or earlier, stimulates discussion as to how different reptile behavior actually is from mammals in reproduction, physiology, sociality, cognition, mentality, and affect. I have addressed these topics previously and in some detail in relation to learning, behavioral development, play, social behavior, physiology, and captive welfare and enrichment (Burghardt, 1977, 1978, 1988, 2013; Burghardt & Layne, 1995). Finally, while elephants are typically considered intelligent and highly social, the extent of their cognitive and social complexity has only recently been recognized and been the focus of scientific study. Similarly, we have greatly underestimated the social and cognitive complexity of reptiles, also little studied; the last decades have led to major findings on reptiles that must be recognized by zoo personnel and incorporated into exhibits and species survival plans (Burghardt, 2013; Burghardt & Milostan, 1995; Doody, Burghardt, & Dinets, 2013; Mendelson, Schuett, & Lawson, 2019)

In this brief chapter, I will make several points based on my presentation and, while discussing some reptile examples, will focus more on embedding work on captive animals in a general behavior systems approach, which formalizes how I think we should approach dealing with captive animals. I apologize for citing many of our lab's publications, as they supplement my brief discussions and provide references to work by dozens of researchers expanding on the points made here and, thus, avoid repetition of theoretical and methodological details. In this way, I hope to provide information of interest to those working on captive animals generally, be they elephants or snakes. Note that the field of animal-human interactions has generally seen large growth as noted in both scholarly treatises (Davis & Balfour, 1992; Freund, McCune, Esposito, Gee, & McCardle, 2016) and more general works (Herzog, 2010). Animal welfare and wellness are now embedded in progressive zoo circles and beginning to include reptiles (Hosey, Melfi, & Pankhurst, 2009; Kaufman, Bashaw, & Maple, 2019; Maple & Perdue, 2013).

APPROACHES TO
LEARNING AND TRAINING

Derived from both operant and Pavlovian conditioning, target train- ing, bridging, and other conditioning methods to safely handle large, potentially dangerous animals are being employed to great effect with elephants and many other large mammals. Regardless of the specific methods used, they can be deployed comparatively in different ways (Burghardt, 2013). These include general process approaches that apply traditional and well-researched phenomena of Pavlovian and instrumen- tal (operant) conditioning, along with habituation and dishabituation, to work with animals to shift enclosures or enter transport caging, stand quietly, accept bodily inspection and medical procedures, and otherwise negotiate captive environments with reduced stress and even enjoyment.

These methods can be used to manage potentially dangerous or dif- ficult reptiles in the zoo, as shown by training a crocodile to allow blood collection (Augustine & Baumer, 2012). Large crocodiles have been shown to be quite receptive to these methods (Augustine, 2009, 2011). Multiple animals in a group setting can also be managed through target training, even snakes such as false water cobras (Gerrits & Augustine, 2013). Target training has been used in many reptiles, and we have used it with monitor lizards, as shown in a Nova documentary, *Lizard Kings*.

More complex cognitive and problem-solving learning methods are more frequent in academic research settings or by academic research- ers in zoos. Often such research, even today, is motivated by the goal of demonstrating that the phyletic position of a species may limit or enable learning more complex tasks. Early approaches were based on the idea that intelligence and cognitive abilities differ greatly across taxonomic categories, and thus, mammals were qualitatively superior to reptiles, reptiles to fish, fish to invertebrates, and so on. Although popular in the past (Bitterman, 1965), the growing field of comparative cognition (Shettleworth, 2010; Vonk & Shackelford, 2012) no longer accepts this type of crude ranking, although some researchers are still interested in analyzing whether certain types of conditioning differ across taxa, such as between rats and toads (Papini & Torres, 2017). In its place is a more neurologically sophisticated approach that uses diverse problem-solving and social tasks and searches for correlations among the size and connections of various parts of the brain subserving these tasks rather than overall brain size or taxonomic position, per se. The extensive literature on cognition in nonhuman primates often focuses on looking for similarities and especially differences across and within primate groups, such as great apes or New World monkeys. Nevertheless, the idea that "lower" vertebrates are less trainable than mammals often taints the attitudes of even those who work with fish, amphibians, and reptiles. There are now many studies showing in both field and captive settings that reptiles have many cognitive abilities that just a few decades ago were considered beyond their station. These include rapid problem box solving in monitor lizards (Manrod, Hartdegen, & Burghardt, 2008) and, among many recent reptile learning studies, response inhibition (Szabo, Noble, & Whiting, 2019).

Ethological and ecological approaches focus on grounding cognitive abilities and training methods in the normal behavior patterns of animals in the environments in which they evolved. The chapter on giraffes by Bashaw in this book provides excellent examples of the consequences of not heeding such environmental constraints as substrate or temperature on behavior and captive welfare.

The ethological approach as reflected in behavior systems theories are particularly applicable when dealing with reptiles, as their behavior, being ectotherms with modest metabolic rates, is often linked quite directly to the environmental challenges affecting their life history and, thus, greatly influence performance of species typical behavior and management procedures.

BEHAVIOR SYSTEMS

All animals have a number of activities that they have to perform in order to survive and reproduce. Among these are foraging and obtaining and processing food and water, locomotion, body care, protection against enemies and weather, negotiating home ranges and territories, competing and sometimes cooperating with conspecifics, courtship and mating, nesting and parental care, social bonding, and others. Often each of these has its own dynamics that, while linked, may affect perceptual, cognitive, and emotional processes in various ways. Learning and experience may alter them in various and complex ways. It is understanding the sequential organization of behavior that is especially important when dealing with conditioning and training of animals for which we have limited information. A major proponent of using behavior systems in animal learning is William Timberlake (1994, 2001), who along with Jerry Hogan (1988, 1994), and Michael Domjan (1994; Domjan, Cusato, & Krause, 2004) developed the conceptual and methodological components of this approach. Recent overviews are in an authoritative textbook (Shettleworth, 2010) and in the recent *APA Handbook of Comparative Psychology* (Call, Burghardt, Pepperberg, Snowdon, & Zentall, 2017), especially in the chapters by Burghardt and Bowers (2017) and Krause and Domjan (2017). Affective/emotional systems are also covered, although not from an explicit behavior systems approach (Panksepp, 2017) in the same compendium, and other useful chapters are to be found there as well. A recent informative historical and conceptual review of behavior systems is also available (Bowers, 2018).

Behavior systems, while overlapping in mechanisms and functions, are often modular in that they may be under the control of discrete sensory, brain, hormonal, and motor systems. There are also large differences among major taxa, and many of these differences relate to sensory abilities, reproductive mode, hatching/birth and neonatal behavior, antipredator adaptations, social organization, and habitat characteristics.

A major feature of behavior systems approaches is that they build on several seminal but still useful ideas from early ethological writers, such as Tinbergen's classic volume on instinct (1951) and Hediger's writings (1950). Knowledge of the basic behavioral repertoire shown by a species in nature, as well as their behavior in captive environments, is an important initial step, and today, the importance of this is widely recognized. The sequential organization of behavior, however, a key component of behavior systems approaches, is less appreciated. The basic picture we owe to the distinctions Wallace Craig developed between appetitive behavior and consummatory acts (1918). The basic model derived from Craig can be depicted in this way:

Need –> drive –> appetitive behavior –> consummatory act –> relative rest (cycling back to drive).

All animals have needs necessary for survival and reproduction. Many, but not all, of these needs result in drives or motivated behavior influenced by both internal states and external stimuli which in turn trigger appetitive behavior directed toward attaining the resource, such as food, mates, or nesting material. Predator or dangerous conspecifics could lead to aversions or avoidance. Appetitive behavior can be quite variable and involve various locomotor, orientation, landmark, and especially learned components. This appetitive phase has more recently been divided in 'general' and 'focal' components hones in on the resource. However, the "goal" for the animal is not necessarily to satisfy the biological or adaptive end so much as it is the attainment of the stimuli allowing the ability to perform the consummatory acts involved in satisfying the need (e.g., killing prey, eating, copulating, playing). These acts are often more stereotyped than behavior seen in the appetitive phases. Performance of the consummatory acts can thus be rewarding in and of themselves, independent of the satisfying of the biological need. Consequently, consummatory acts reduce the drive, leading to relative rest, while the associated need may be satisfied in a more delayed mode (e.g., sugar and sodium changes in blood for hunger and thirst). Understanding the organization of the behavior we encounter with our captive animals can best target those aspects that can be modified or controlled most easily. Target training, for example, usually focuses on a consummatory act reward, such as food or even touching and stroking affectionately. When the drive is reduced, the animal returns to the relative rest stage, reflecting the reduction of the motivation or mood and, indirectly, the underlying need. In actuality, several such Craig-style systems may operate simultaneously or even in conflict with one another, as when an animal may show interest in food presented by a keeper and yet also express interest in retiring to a secluded area away from visitors or other distracting stimuli. Experienced keepers can "read" these tendencies, especially if very familiar with the individuals. Efforts should be made to identify what these cues are down to specific postures, facial expressions, ear and tail positions, sounds, and so on.

CRITICAL ANTHROPOMORPHISM

In studying and describing behavior systems, and animal behavior in general, it is important to use objective terms, and if more loaded ones are used such as wanting, liking, fearing, and so on, these should be based on a critical, not uncritical, anthropomorphism. This is especially applicable to the zoo environment, where exotic species are sometimes maintained and managed in spite of a lack of detailed knowledge about their behavior in the field or successful experience with them in captivity. Critical anthropomorphism is combining our intuitions and the way we might behave in a comparable situation (e.g., obtaining hidden food or escaping the pen) with the best available scientific knowledge about the sensory, motor, social, antipredator, dietary, habitat, and other aspects of the species' normal behavior (Burghardt, 1998). In the rush to avoid anthropomorphic terms, we can actually ignore the animal's perspective, or anthropomorphism by omission (Rivas & Burghardt, 2002). Such problems arise with unfamiliar animals, as are many reptiles, exotic animals generally, and especially rare and endangered species.

ENVIRONMENTAL AND
BEHAVIORAL ENRICHMENT

In the early years of zoo biology (actually not that long ago), zoos were often led by directors and others with limited behavioral understanding, but with a strong focus on physical health and "clean" enclosures. This led to the tyranny of the additive model (Burghardt, 1996, 1998), in which investment in naturalistic enclosures and encouraging species-typical behavior took a back seat to veterinary concerns. This is perhaps understandable, as historically veterinarians directed many zoos, and some do still. Providing for the behavioral needs of animals cost money and keeper time and effort and, thus, was considered more of a nice luxury, if it could be afforded, but of secondary importance. This attitude has changed in many zoos over the last decades (Maple, 1996), as shown by all the zoos represented in this volume, though the goalposts should, and are, constantly moving, and financial considerations are ever present. Environmental enrichment is now enshrined in many zoos, most typically with mammals, but reptiles are still underrepresented. Monitor lizards, rightly reputed for intelligence and being active and dexterous, were the first to be enriched. Enrichment could be simply dispersing food around the enclosure rather than just in one spot. This is relatively easy to implement, inexpensive, and makes for more diverse and active behavior that benefits the animals as well as the visitors viewing them (Markowitz & Gavazzi, 1996). However, enrichment can also deter boredom, foster problem solving, or enhance prey handling and other tasks, as we showed in studies of turtles, monitors, and snakes, respectively (Almli & Burghardt, 2006; Burghardt et al., 2002; Burghardt, Ward, & Rosccoe, 1996; Manrod, Hartdegen, & Burghardt, 2008) and which may also involve play. New studies of enrichment (e.g., Bashaw, Gibson, Schowe, & Kucher, 2016) and complex learning in reptiles are appearing in recent years, a most welcome development. A few sources for such studies are listed in reviews (Burghardt, 2013; Wilkinson & Huber, 2012) and include social learning as well as individual cognitive feats. A recent zoo-based study on the common North American box turtle is an exemplary model of what can be shown in reptiles using methods pioneered with mammals and birds (Leighty, Grand, Pittman Courte, Maloney, & Bettinger, 2013). With the emphasis on captive breeding of reptiles in zoos, a cautionary paper on the effects of incubation temperature on learning in lizards is relevant (Clark, Amiel, Shine, Noble, & Whiting, 2014). All this is to suggest that reptiles are now being viewed as alert cognitive actors rather than small-brained automatons. Zoos and their staff need to embrace enhancing their lives with a passion equal to that expended on other animals and alter biases against them, especially snakes (Burghardt, Murphy, Chiszar, & Hutchins, 2009).

SOME FINAL CAUTIONARY NOTES

I have argued that the term enrichment, either environmental or behavioral, is in some ways unfortunate, as it implies that we are doing something beyond the normal for the species involved (Burghardt, 1996). In point of fact, all our animals are in a state of deprivation in the sense that many natural stimuli are lacking, be they habitat complexity, diet, space, seasonality, social groupings, climatic variability, and, most strikingly, predators. Of course, captive animals may also avoid diseases, illnesses, conspecific aggression, and food stresses they might normally encounter. Still, the fact remains that all captive wild species are in a state of controlled deprivation. Our goal is to identify those components of such deprivation that facilitate or inhibit the occurrence of naturalistic behavior and increase the former and eliminate or reduce the latter. This, of course, is no mean task and especially true for large far-ranging animals such as elephants and giraffes as well as many predators. These are also concerns with large reptiles.

An additional factor is the recognition that animals are individuals with their own personalities and attachments to conspecifics and keepers. This is well appreciated with many mammals and especially elephants. A growing literature is documenting similar phenomena in many reptiles (Waters, Bowers, & Burghardt, 2017). So here again there are useful connections between those who study elephants and croco-diles! With large, potentially dangerous animals, it is especially critical that the behavior systems of animals and their sequential organization be analyzed and integrated into zoo practice so that our management toolbox is effectively deployed for the benefit of both the keeper and the kept.

REFERENCES

Almli, L., & Burghardt, G. M. (2006). Environmental enrichment alters the behavioral profile of ratsnakes (*Elaphe*). *Journal of Applied Animal Welfare Science, 9*, 85–109.

Augustine, L. (2009). Husbandry training with an exceptional South African crocodile. *ABMA Wellspring, 10*(3), 2–3.

Augustine, L. (2011). Putting training to work in a large animal capture. *ABMA Wellspring, 12*(1–2), 36–37.

Augustine, L., & Baumer, M. (2012). Training a Nile crocodile to allow for collection of blood at the Wildlife Conservation Society's Bronx Zoo. *Herpetological Review, 43*, 432–435.

Bashaw, M. J., Gibson, M. D., Schowe, D. M., Kucher, A. S. (2016). Does enrichment improve reptile welfare? Leopard geckos (*Eublepharis macularius*) respond to five types of environmental enrichment. *Applied Animal Behaviour Science, 184*, 150-160.

Bitterman, M. E. (1965). Phyletic differences in learning. *American Psychologist, 20*, 396–410.

Bowers, R. I. (2018). A common heritage of behaviour systems. *Behaviour, 155*, 415-442.

Burghardt, G. M. (1977). Learning processes in reptiles. In C. Gans & D. Tinkle (Eds.), *The biology of reptilia: Ecology and behavior* (Vol. 7) (pp. 555–681). New York, NY: Academic Press.

Burghardt, G. M. (1978). Behavioral ontogeny in reptiles: Whence, whither, and why. In G. M. Burghardt & M. Bekoff (Eds.), *The development of behavior: Comparative and evolutionary aspects* (pp. 149–174). New York, NY: Garland STPM Press.

Burghardt, G. M. (1988). Precocity, play, and the ectotherm-endotherm transition: Superficial adaptation or profound reorganization? In E. M. Blass (Ed.), *Handbook of behavioral neurobiology* (Vol. 9) (pp. 107–148). New York, NY: Plenum.

Burghardt, G. M. (1996). Environmental enrichment or controlled deprivation? In G. M. Burghardt, J. T. Bielitski, J. R. Boyce, & D. O. Schaefer (Eds.), *The well-being of animals in zoo and aquarium-sponsored research* (pp. 91–101). Greenbelt, MD: Scientists Center for Animal Welfare.

Burghardt, G. M. (1998). Snake stories: From the additive model to ethology's fifth aim. In L. Hart (Ed.), *Responsible conduct of research in animal behavior* (pp. 77–95). United Kingdom: Oxford University Press.

Burghardt, G. M. (2013). Environmental enrichment and cognitive complexity in reptiles and amphibians: Concepts, review, and implications for captive populations. *Applied Animal Behavior Science, 147,* 286–298.

Burghardt, G. M., & Bowers, R. I. (2017). From instinct to behavior systems: An integrated approach to ethological psychology. In J. Call, G. M. Burghardt, I. M. Pepperberg, C. T. Snowdon, & T. Zentall (Eds.), *APA handbook of comparative psychology: Basic concepts, methods, neural substrate, and behavior* (Vol. 1) (pp. 333–364). Washington, DC: American Psychological Association.

Burghardt, G. M., Chiszar, D., Murphy, J. B., Romano, J., Walsh, T., & Manrod, J. (2002). Behavioral diversity, complexity, and play in Komodo Dragons. In J. B. Murphy, C. Ciofi, C. de la Panouse, & T. Walsh (Eds.), *Komodo dragons: Biology and conservation*, (pp. 78–117). Washington, DC: Smithsonian Press.

Burghardt, G. M., & Layne, D. G. (1995). Effects of ontogenetic processes and rearing conditions. In C. Warwick, F. L. Frye, & J. B. Murphy (Eds.), *Health and welfare of captive reptiles* (pp. 165–185). London, United Kingdom: Chapman & Hall.

Burghardt, G. M. & Milostan, M. Ethological studies on reptiles and amphibians: Lessons for species survival. In *Captive conservation of endangered species: An interdisciplinary approach.* (J. Demarest, B. Durrant, & E. Gibbons, eds.). SUNY Press, New York, 1995, 187-203.

Burghardt, G. M., Murphy, J. B., Chiszar, D., & Hutchins, M. (2009). Combatting ophiophobia: Origins, treatment, education, and conservation tools. In S. J. Mullin & R. A. Seigel (Eds.), *Snakes: Ecology and conservation* (pp. 262–280). Ithaca, NY: Comstock.

Burghardt, G. M., Ward, B., & Rosccoe, R. (1996). Problem of reptile play: Environmental enrichment and play behavior in a captive Nile soft-shelled turtle (*Trionyx triunguis*). *Zoo Biology, 15,* 223–238.

Call, J., Burghardt, G. M., Pepperberg, I. M., Snowdon, C. T., & Zentall, T. (Eds.). (2017). *APA handbook of comparative psychology* (Vols. 1 & 2). Washington, DC: American Psychological Association.

Clark, B. F., Amiel, J. J., Shine, R., Noble, D. W. A., & Whiting, M. J. (2014). Color discrimination and associative learning in hatchling lizards incubated at "hot" and "cold" temperatures. *Behavioral Ecology and Sociobiology, 68*, 239–247.

Craig, W. (1918). Appetites and aversions as constituents of instincts. *Biological Bulletin, 34*, 91–107.

Davis, H., & Balfour, D. (Eds.). (1992). *The inevitable bond: Examining scientist-animal interactions*. United Kingdom: Cambridge University Press.

Domjan, M. (1994). Formulation of a behavior system for sexual conditioning. *Psychonomic Bulletin & Review, 1*, 421–428.

Domjan, M., Cusato, B., & Krause, M. (2004). Learning with arbitrary versus ecological conditioned stimuli: Evidence from sexual conditioning. *Psychonomic Bulletin & Review, 11*, 232–246.

Doody, J. S., Burghardt, G. M., & Dinets, V. (2013). Breaking the social-nonsocial dichotomy: A role for reptiles in vertebrate social behavior research? *Ethology, 119*, 95–103.

Freund, L. S., McCune, S., Esposito, L., Gee, N. R., & McCardle, P. (Eds.). (2016). *The social neuroscience of human-animal interaction*. Washington, DC: American Psychological Association.

Gerrits, J., & Augustine, L. (2013). Multiple snakes, multiple problems. *AAZK Animal Keeper's Forum, 40*(7), 318–319.

Hediger, H. (1950). *Wild animals in captivity*. London, United Kingdom: Butterworth.

Herzog, H. (2010). *Some we love, some we hate, some we eat: Why it's so hard to think straight about animals*. New York, NY: Harper Collins.

Hogan, J. A. (1988). Cause and function in the development of behavior systems. In E. M. Blass (Ed.), *Handbook of behavioral neurobiology* (Vol. 9) (pp. 63–106). New York, NY: Plenum.

Hogan, J. A. (1994). Structure and development of behavior systems. *Psychonomic Bulletin & Review, 1*, 439–450.

Hosey, G., Melfi, V., & Pankhurst, S. (2009). *Zoo animals: Behavior, management, and welfare*. United Kingdom: Oxford University Press.

Kaufman, A. B., Bashaw, M. J., & Maple, T. L. (Eds.) (2019). *Scientific foundations of zoos and aquariums: Their role in conservation and research*. Cambridge, UK: Cambridge University Press.

Krause, M. A., & Domjan, M. (2017). Ethological and evolutionary perspectives on Pavlovian conditioning. In J. Call, G. M. Burghardt, I. M. Pepperberg, C. T. Snowdon, & T. Zentall (Eds.), *APA handbook of comparative psychology: Perception, learning, and cognition* (Vol. 2) (pp. 247–256). Washington, DC: American Psychological Association.

Leighty, K. A., Grand, A. P., Pittman Courte, V. L., Maloney, M. A., & Bettinger, T. L. (2013). Relational responding by eastern box turtles (*Terrepene carolina*) in a series of color discrimination tasks. *Journal of Comparative Psychology, 127*, 256–254.

Manrod, J., Hartdegen, R., & Burghardt, G. M. (2008). Rapid solving of a problem apparatus by juvenile black-throated monitor lizards (*Varanus albigularis albigularis*). *Animal Cognition, 11*, 267–263.

Maple, T. L. (1996). Introductory remarks: The art and science of enrichment. In G. M. Burghardt, J. T. Bielitski, J. R. Boyce, & D. O. Schaefer (Eds.), *The well-being of animals in zoo and aquarium-sponsored research* (pp. 79–84). Greenbelt, MD: Scientists Center for Animal Welfare.

Maple, T. L., & Perdue, B. M. (2013). *Zoo animal welfare*. Berlin, Germany: Springer-Verlag.

Markowitz, H., & Gavazzi, A. J. (1996). Definitions and goals of enrichment. In G. M. Burghardt, J. T. Bielitski, J. R. Boyce, & D. O. Schaefer (Eds.), *The well-being of animals in zoo and aquarium-sponsored research* (pp. 5–90). Greenbelt, MD: Scientists Center for Animal Welfare.

Mendelson, III, J. R., Schuett, G. W., & Lawson, D. P. (2019). Krogh's principle and why the modern zoo is important to academic research. In A. B. Kaufman, M. J. Bashaw, and T. L. Maple (Eds.), *Scientific foundations of zoos and aquariums: Their role in conservation and research* (pp. 586-617). Cambridge, UK: Cambridge University Press.

Panksepp, J. (2017). Instinctual foundations of animal minds: Comparative perspectives on the evolved affective neural substrate of emotions and learned behaviors. In J. Call, G. M. Burghardt, I. M. Pepperberg, C. T. Snowdon, & T. Zentall (Eds.), *APA handbook of comparative psychology: Basic concepts, methods, neural substrate, and behavior* (Vol. 1) (pp. 475–500). Washington, DC: American Psychological Association.

Papini, M. R., & Torres, C. (2017). Comparative learning and evolution. In J. Call, G. M. Burghardt, I. M. Pepperberg, C. T. Snowdon, & T. Zentall (Eds.), *APA handbook of comparative psychology: Perception, learning, and cognition* (Vol. 2) (pp. 267–286). Washington, DC: American Psychological Association.

Rivas, J., & Burghardt, G. M. (2002). Crotalomorphism: A metaphor for understanding anthropomorphism by omission. In M. Bekoff, C. Allen, & G. M. Burghardt (Eds.), *The cognitive animal: Theoretical, methodological, and empirical approaches* (pp. 9-17). Cambridge, MA: MIT Press.

Shettleworth, S. J. (2010). *Cognition, evolution, and behavior* (2nd ed.). United Kingdom: Oxford University Press.

Szabo, B., Noble, D. W. A., Whiting, M. J. (2019). Context-specific response inhibition and differential impact of a learning bias in a lizard. *Animal Cognition, 22,* 317-329.

Timberlake, W. (1994). Behavior systems, associationism, and Pavlovian conditioning. *Psychonomic Bulletin & Review, 1,* 405–420.

Timberlake, W. (2001). Motivational modes in behavior systems. In R. R. Mowrer & S. B. Klein (Eds.), *Handbook of contemporary learning theories* (pp. 155–209). Mahwah, NJ: Lawrence Erlbaum Associates.

Tinbergen, N. (1951). *The study of instinct.* Oxford, United Kingdom: Clarendon Press.

Vonk, J., & Shackelford, T. K. (Eds.). (2012). *The Oxford handbook of comparative evolutionary psychology.* Oxford, UK: Oxford University Press.

Waters, R. M., Bowers, B. B., & Burghardt, G. M. (2017). Personality and individuality in reptile behavior. In J. Vonk, A. Weiss, & S. A. Kuczaj (Eds.), *Personality in nonhuman animals*, (pp. 153–184). New York, NY: Springer.

Wilkinson, A., & Huber, L. (2012). Cold-blooded cognition: Reptilian cognitive abilities. In J. Vonk & T. K. Shackelford, *The Oxford handbook of comparative evolutionary psychology* (pp. 129–143). Oxford, UK: Oxford University Press.

Chapter 17

TOWARD A WELLNESS-INSPIRED EXHIBIT FOR ELEPHANTS AT THE JACKSONVILLE ZOO AND GARDENS

Terry L. Maple, Jeff Sawyer, Gary H. Lee, J. Hanuliakova,
and Dan Maloney
Jacksonville Zoo and Gardens, CLR Design

For too many years, zoo elephants throughout the world have been neglected by our failure to act on our accumulated knowledge of elephant behavior (Kane, 2009; Maple, Bloomsmith, & Martin, 2009). Field and captive studies by zoologists and psychologists continue to identify key environmental and social factors that contribute to the psychological well-being (wellness) of elephants living in zoological parks and in nature (Berg, 1983; Langbauer, 2000; Lee & Moss, 1999; Moss & Poole, 1983; Poole, 1994; Poole & Moss, 2008; Shoshani & Eisenberg, 1992; Viljoen, 1989; Wyat & Eltringham, 1974). There may be good reasons why it took so long to design and build better elephant exhibits, but ignorance isn't one of them. Fortunately, a new generation of elephant exhibits have appeared in North America and around the world. These modern facilities have evolved to represent the best ideas and interpretations of field scientists, educators, and zoo biologists.

The world's largest land mammal, elephants evolved to live in large herds led by mature females whose cognitive capacity enabled the herd to negotiate complex and ever-changing habitats. They roamed widely to find food, water, and join with other familiar elephants remembered by the long-lived matriarchs. The complex life of elephants in Africa and

Asia requires the intellect and the stamina to outthink and outrun heavily armed poachers who hunt them relentlessly for their valuable ivory tusks. By nature, elephants are resilient with superior mental capacity and physical strength. It is a challenge to meet the needs of such complex creatures, but we are obligated to exercise our creativity on their behalf. Only when we meet their needs can we justify keeping them in a zoo.

We have also discovered how to improve the living conditions of elephants in substandard facilities, but these temporary fixes must eventually lead to comprehensive, permanent changes (Brockett, Stoinski, Black, Markowitz, & Maple, 1999; Wilson et al., 2006). Thankfully, there are fewer and fewer inferior facilities in the world's accredited zoos. To achieve a higher quality of life, zoo administrators must be prepared to provide appropriate facilities that replicate in form and function an elephant's natural habitat. A blueprint for an optimal elephant exhibit was envisioned by David Hancocks (2009), who identified an array of options that future zoos can choose to enrich the lives of zoo elephants and the people who want to see them thriving like their wild kin.

An investment in elephants requires a lifelong commitment to the animals and their offspring, and an exhibit that must be able to expand to accommodate a growing herd of males and females. Inspired by the words of our colleagues at the Jacksonville wellness workshop, we have embraced the challenge of designing, building, and operating a superior elephant exhibit (see Figure 1). To reach this goal, we are working to acquire additional space on the edges of the zoo property. Optimal elephant exhibits are among the most expensive facilities a zoo will ever build on its campus, so the Jacksonville Zoo exhibit will require significant funding. In a few years, once opened and operational, our exhibit will take its place among other creative exhibits worldwide to ensure that every elephant in managed care is thriving. *Wellness for Elephants* has been published and disseminated to encourage collaboration in designing and ultimately building facilities that reach the highest standards and best practices in the exhibition and management of African and Asian elephants.

FIGURE 1. OPPORTUNITIES FOR EXPANDING THE ELEPHANT EXHIBIT ON
THE JACKSONVILLE ZOO PROPERTY. (COURTESY CLR DESIGN).

ELEPHANTS LIVING LARGE

Theoretically, herd size is an important contributor to individual wellness. Although there are limits to the size of zoo herds due to availability and compatibility, larger herds of seven to 10 individuals encourage
the social complexity and opportunity that elephants need. Although
no minimal sizes have been specified by experts, larger is clearly better.
Integration of males and females is possible in the zoo if conditions and
facilities are designed for their group management. No elephant should
live in isolation. To properly manage social herds and social opportunity,
an exhibit must be carefully engineered to accommodate activity and
exploration, reproduction, socialization, autonomy, and cognitive complexity. To replicate nature, zoo herds must be structured for stability. To
achieve these positive outcomes, zoo managers must give priority to social structure and group cohesion. Although translocations are stressful

and may disrupt stability, the formation of larger groups will require the careful movement of elephants from one location to another.

Some elephant herds in North America have been recently expanded with the addition of animals imported from the wild. It appears that these animals have adjusted well, but it is uncertain how many more elephants will be acquired in this way. Expanding herds to improve animal welfare is necessary, but reproduction is opposed by extremists in the animal rights movement. However, as superior facilities become the norm, elephants born into managed care will reap the benefits of innovative, naturalistic habitats tailored to their every need. These exhibits cannot be static; they must continue to evolve and improve, and our visitors must be educated to understand that elephants can live well in state-of-the art zoo exhibits. To retain and expand the support of our communities, we must be willing to actively debate our critics and market our achievements in animal welfare. Animals born and raised by their mothers in normal social groups in naturalistic zoos will not suffer; they will thrive.

Four acres is the minimum amount of space that elite zoos are currently devoting to superior elephant exhibits, but some are much larger. The Oklahoma City Zoo recently constructed a 9.5-acre exhibit for the Asian species, currently the largest North American zoo exhibit for elephants. In Jacksonville, we aim to give them much more than the minimum of 4 acres with the hope of distributing the animals within two interconnected forested areas. The total acreage available for Jacksonville's African herd could be as much as 15 acres. Living space will be configured so elephants are encouraged to walk at least 5 miles per day. Others (e.g., Barber, 2009) have recommended a standard of 7 miles. Female African herds in the wild have been recorded by Wyat and Eltringham (1974) walking at a pace of 9 miles per day. While there is no magic number, optimal health requires sufficient acreage to take a long walk daily. Walking and running promotes foot health in elephants and should be encouraged by keepers. Some zoos encourage walking by scheduled daily training sessions. Activation is one of the benefits of the human-animal relationship, and sophisticated training techniques enable activation and husbandry.

At the Jacksonville Zoo and Gardens, we are planning a walking trail for elephants that supplements the exhibit space and provides a protected perimeter to take the animals "on safari" within view of our guests (Figure 2). Ideally, the trail will connect the two domains of forest devoted to elephants. This trail resembles in concept the Buckhead Elephant Park diagram depicted in chapter 10 in the book *An Elephant in the Room* (Maple, Bloomsmith, & Martin, 2009). From this perspective, zoo visitors will learn about the nomadic nature of elephant life in the wild. The Jacksonville elephants will be able to walk to a location where they can take a dip in the adjacent Trout River. This will likely be a simulated dip in a manufactured full-immersion pool with an infinite edge indistinguishable from the actual river boundary. Elephants love to play and swim in water, and our guests will surely enjoy the moments when our elephants frolic and interact in our simulated river. As the elephants walk to their riverine destination, it will appear to visitors that they are immersed among the elephants who will appear along the trail in new locations in the zoo.

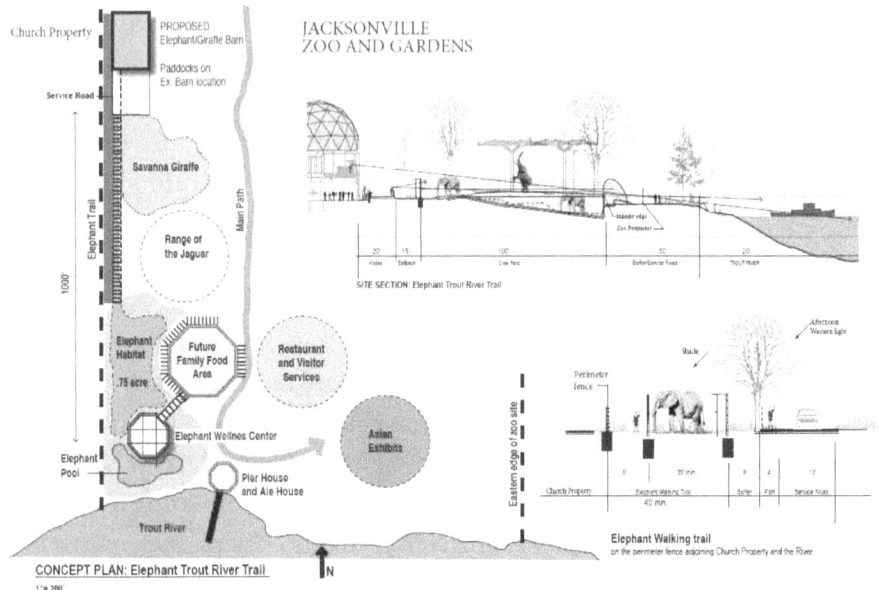

FIGURE 2. SAFARI WALK TO TROUT RIVER IMMERSION POOL FROM
A CONCEPT BY NEVIN LASH (URSA INTERNATIONAL).

A world-class elephant exhibit must be large enough and complex enough for a group of elephants and their offspring, but acreage is a simplified way to calculate space in the zoo. In fact, exhibit space can be broken down into components and usability. As Hediger (1969) confirmed, topography is very important to large animals, especially elephants. Modern zoos have abandoned hard surfaces in night barns in favor of soft substrates of sand and soil. Softer substrates help to activate elephants by easing the pain on their joints. Objects in the environment, simulated trees and the like, are there to provide rubbing surfaces and manipulatable toys. Bulls, for example, enjoy pushing against a moveable rock or ball. Immersive pools and mudholes when available are used daily by elephants. They also benefit from water sources that can be structured to provide a useable shower or waterfall and spouts that can be operated by keepers to squirt the animals. We are also committed to engineering technology that enables the elephants to squirt water on one another and possibly engage in playful squirts in the direction of zoo visitors. All of these features will be developed and tested in Jacksonville. Because keepers no longer go into the enclosure with elephants, one safe way to playfully interact with them is to provoke friendly water fights from a safe distance.

Because elephants are also tall animals with long, flexible trunks, they can reach high into the canopy of trees in the wild. We are planning to provide vertical opportunities for the Jacksonville elephants with the construction of feeding platforms that provide access to food that the elephants have to stretch to obtain. In exhibits that provide verticality, we prefer to describe living space in terms of cubic volume. For many animals, particularly arboreal primates, it is more informative to design exhibits that can be evaluated in terms of their volumetric and surface utility. Elephants don't climb trees or walls, but they probe them with their trunks. Therefore, we must accommodate their abilities and propensities by building up. We will design our habitats to provide as much useable volume as the world's best zoo exhibits for elephants.

TO ENABLE THRIVING

To reach the high standards that promote thriving, elephants must be able to exercise the full behavioral repertoire of wild elephants. Working with the world's most experienced elephant exhibit designers, CLR of Philadelphia, we are assembling the best ideas to create an exhibit without peer. With over ten previous elephant exhibits designed by CLR for North American zoos, no innovation will be left out of the Jacksonville exhibit. We will provide for mental stimulation designed by psychologists who understand the full mental capacity of this charismatic species. To ensure that the Jacksonville elephants thrive, we will pay special attention to mental and physical challenges, one of the most important factors in Maslow's (1962) recommendations for achieving "self-actualization" (see chapter 1). For example, we can provide elephants with a cognitive problem that can be solved via computer interface designed for an elephant's trunk. Once the problem is solved, the elephants can activate the door leading to the safari pathway to the Trout River bathing site. In this way, the elephants are stimulated mentally and physically and rewarded for their efforts. Our first application of this technology is contained within the architecture of the new African Forest Exhibit, opened in September 2018. The focal point of this new exhibit is the "Wellness Tree," designed to provide for animal-computer interactions within view of the public. With this technology, we will challenge bonobos, gorillas, and mandrills to solve cognitive problems, thereby enriching their zoo experience and learning more about their mental capacity. The logical framework for this process was introduced through Julia Hanuliakova's reproduction of A. H. Maslow's Hierarchy of Needs (Figure 3). In this diagram, the highest needs (choice, social and mental stimulation) lead to wellness.

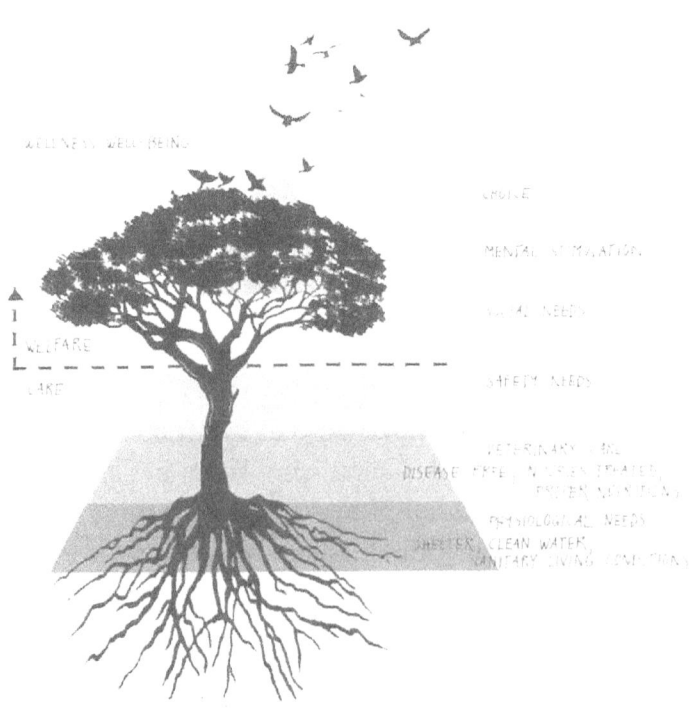

FIGURE 3: A.H. MASLOW'S HIERARCHY OF NEEDS.

We will devise other opportunities that return environmental control to the elephants. They will be able to control lighting in the barn, set the temperature (within limits), engage naturalistic sounds such as elephant vocalizations or forest insects and birds and infrasonic sound stimuli that only elephants can hear, activate aromas at random sites within their enclosures, operate feeding stations and food puzzles, and obtain visual access to their keepers working or volunteers engaged in food preparation at a remote site. We will brainstorm as long as it takes to keep the ambient environment refreshed and utterly novel. These technical innovations will encourage zoo elephants to behave like wild elephants. Of course, the most stimulating variables are social variables enabled by the formation of social groups. Daily access to other elephants is the most important opportunity our new exhibit will provide. Borrowing from the Dallas Zoo motto, Jacksonville Zoo designers intend to "let elephants be elephants" in every possible way.

AESTHETICS AND LANDSCAPE

The new Jacksonville Zoo and Gardens' master plan provides a beautiful naturalistic setting for the elephant herd where zoo visitors will gain respect for the unique adaptations of these animals. The Jacksonville Zoo campus is known for its aesthetics and its institutional commitment to botanicals. Because the elephants will be able to express their full repertoire of behavior, we expect them to be active and curious throughout the day and night. We will be sure to provide information through sophisticated graphics and scheduled conversations with keepers and educators to help our guests participate in the elephant conservation programs that we support. Our exhibit will be designed to educate and inspire all of our visitors and encourage them to learn more on our website and in social media postings as follow-ups to a zoo visit. The landscape will be as authentic as possible while framing the animals in a background reminiscent of a photographic safari to Africa. The zoo design team will make it easy to link the zoo to endangered wild habitats in need of external funding and expertise. Innovations of this type were foreseen by David Hancocks (2009), who described visionary zoos of the future:

> At last, almost 200 years after the first modern public zoological garden opened its gates in London's Regents park in 1828, zoos in wealthy countries are beginning to serve as effective bridges between their visitors and the conservation of wild places. In this second decade of the 21st century, zoos are serving as links between the two conflicting major clients—human audiences from urban environments, and wild animals from wilderness habitats—both metaphorically and in practical ways. They have finally graduated to a position where their existence is justified, their philosophy is meaningful, and the purpose is truly valuable.

EVIDENCE-BASED DESIGN

Dedicated zoo professionals at the Jacksonville Zoo and Gardens are committed to helping others to design better exhibits by engaging in the serious study of our own elephants and the unique exhibit designed for them. Deploying students and staff, under the supervision of our wellness research professionals and academic research partners, we will continue to gather data and publish in reputable professional journals. Periodic meetings similar to the Wellness for Elephants workshop summarized in this book will continue to drive innovation, and we will actively disseminate information on behalf of elephants and other zoo animals. A post-occupancy evaluation (POE) of the new exhibit is a high priority once it has been completed. Each time the exhibit experiences significant physical change, we will objectively evaluate the effects on the animals, staff, and visitors. The first author's research team has years of experience conducting and publishing evaluations of this kind (Maple & Finlay, 1986; Ogden, Finlay, & Maple, 1990; Hoff, Forthman, & Maple,

1994; Hoff & Maple, 1995; Chang, Forthman, & Maple, 1999; Stoinski, Hoff, & Maple, 2001; Lukas, Hoff, & Maple, 2003; Wilson, Kelling, Poline, Bloomsmith, & Maple, 2003; Maple, 2005; Bashaw, Kelling, Bloomsmith, & Maple, 2007).

When the new Jacksonville elephant exhibit is fully operational, we anticipate that Jacksonville Zoo and Gardens will take its place as a global intellectual center for the study of elephant biology, behavior, design, and wellness. We are committed to recruiting additional staff to drive this empirical vision and to engage with conservation partners who are working to protect elephants in the natural world. In addition, elephant experts will be periodically invited to speak at public forums to celebrate our significant investment in elephant conservation and welfare. By visiting with experts, we will provide continuing education for our staff while enlightening our visitors and the Jacksonville community.

The first steps toward an empirical elephant center have already been taken. A wildlife wellness department was established at the Jacksonville Zoo and Gardens on June 1, 2014. With the recruitment of the first applied animal behavior analyst, Valerie Segura, the department began to organize a wellness services program and research lab comprised of institutional leadership, dedicated staff, internal zoo collaborators, and visiting students from nearby universities. Our research and services team has been supported by a percentage of zoo admissions revenue and an anonymous donor's financial commitment of more than $1 million over seven years. This same donor also made a $1 million gift to support the enrichment and research technology at the African Forest Exhibit. Thus far, we have sponsored three master's-level graduate students, one of whom (Kaylin Tennant) recently entered the doctoral program at Case Western Reserve University affiliated with Cleveland Metroparks Zoo. She defended her master's thesis on December 1, 2017, with data she gathered on multimale groups of gorillas in two zoos. Since its inception, the Jacksonville wellness group and our collaborators at the University of Florida and the University of North Florida have published 20 peer-reviewed papers and book chapters and delivered many presentations at national and international meetings. Many other papers are in preparation for publication. *Wellness for Elephants* is the latest project managed through the publication process by our entire wellness team.

TOWARD A GLOBAL WILDLIFE WELLNESS NETWORK

The Wellness for Elephants workshop was a first opportunity to gather zoo professionals and academic colleagues in Jacksonville to share information on how to upgrade elephant management and design superior exhibits. This is just the beginning of a long-term commitment to improve quality of life for all animals at the Jacksonville Zoo and Gardens while assisting other institutions with the same goals. Hopefully, the contributions in this special volume will lead to new and better ideas and a stronger commitment to facilities that bring out the best in the animals who reside within. While wellness is facilitated by astute zoo designers who understand animal behavior, even the best exhibits will not succeed unless they are managed properly by thoughtful animal managers, veterinarians, and caregivers. Our workshop recruited professionals who have designed excellent exhibits and those who have been innovators in managing them. The psychology of elephants has been revealed in the papers we have studied and the ideas expressed at our workshop. Due to the urgency of elephant reforms, the literature on elephant behavior and elephant welfare must continue to expand.

After decades of working to advance the science and practice of animal welfare, it is our belief that the discoveries emanating from a network of empirical zoos will inevitably lead to the correct formula for a given species to live well in managed care. To ensure this outcome, we must fully understand the psychology of each species and each and every individual in the zoo population. The thoughtful contributions in *Wellness for Elephants* represent a catalogue of proven ideas and promising innovations from experts who offer this material to benefit the animals, the employees devoted to their health and well-being, and the people who spend their time and money to support zoos of quality and consequence. Copies of *Wellness for Elephants* are available at Amazon. We've made it available at a modest price in the hopes that zoo keepers, docents, students, and all members of the zoo community will be able to purchase the book. On behalf of our colleagues and friends at the Jacksonville Zoo and Gardens, we hope you will enjoy the book and put it to good use.

REFERENCES

Barber, J. C. E. (2009). Unpacking the trunk: Using basic research approaches to identify and address captive elephant welfare concerns. In D. L. Forthman, L. F. Kane, D. Hancocks, & P. F. Waldau (Eds.), *An elephant in the room: The science and well-being of elephants in captivity* (pp. 181–188). North Grafton, MA: Tufts University Press.

Bashaw, M. J., Kelling, A. S., Bloomsmith, M. A., & Maple, T. L. (2007). Environmental effects on the behavior of zoo-housed lions and tigers with a case study of the effects of a visual barrier on pacing. *Journal of Applied Animal Welfare Science,* 10(2), 95–109.

Berg, J. (1983). Vocalizations and associated behaviours of the African elephant (*Loxodonta africana*) in captivity. *Zeitschrift fur Tierpsychologie, 63,* 63–79.

Brockett, R., Stoinski, T., Black, J., Markowitz, T., & Maple, T. L. (1999). Nocturnal behavior in a group of unchained African elephants. *Zoo Biology, 18*(2), 101–109.

Chang, T. R., Forthman, D. L., & Maple, T. L. (1999). Comparison of captive mandrills in traditional and ecologically representative exhibits. *Zoo Biology, 18*(3), 163–176.

Hancocks, D. What will new zoo environments look like? (2009). In D. L. Forthman, L. F. Kane, D. Hancocks, & P. F. Waldau (Eds.), *An elephant in the room: The science and well-being of elephants in captivity* (pp. 215-225). North Grafton, MA: Tufts University Press.

Hediger, H. (1969). *Man and animal in the zoo*. New York, NY: Delacorte Press.

Hoff, M. P., Forthman, D., & Maple, T. L. (1994). Dyadic interactions of infant lowland gorillas in an outdoor exhibit compared to an indoor holding area. *Zoo Biology, 13*(3), 245–256.

Hoff, M. P., & Maple, T. L. (1995). Post-occupancy modification of a lowland gorilla enclosure at Zoo Atlanta. *International Zoo Yearbook, 34*, 153–160.

Langbauer, W. R., Jr. (2000). Elephant communication. *Zoo Biology, 19*, 425–445.

Lee, P. C., & Moss, C. J. (1999). The social context for learning and behavioral development among wild African elephants. In H. L. Box & K. R. Gibson (Eds.), *Mammalian social learning* (pp. 102-105). United Kingdom: Cambridge University Press.

Lukas, K. E., Hoff, M. P., & Maple, T. L. (2003). Gorilla behavior in response to systematic alternation between zoo enclosures. *Applied Animal Behavior Science, 81*(4), 367–386.

Maple, T. L. (2005). Post-occupancy evaluation in the zoo: Toward a science of appropriate, functional, and superior exhibitry for animals and people. In A. B. Plowman & S. J. Tonge (Eds.), *Innovation or replication: Proceedings of the Sixth International Symposium on Zoo Design* (pp. 111–117). Paignton, Devon, United Kingdom: Whitley Wildlife Preservation Trust.

Maple, T. L., Bloomsmith, M. A., & Martin, A. (2009). Primates and pachyderms: A primate model of zoo elephant welfare. In D. L. Forthman, L. F. Kane, D. Hancocks, & P. F. Waldau (Eds.), *An elephant in the room: The science and well-being of elephants in captivity* (pp. 129–153). North Grafton, MA: Tufts University Press.

Maple, T. L., & Finlay, T. W. (1986). Evaluating the environments of captive nonhuman primates. In K. Benirschke (Ed.), *Primates: The road to self-sustaining populations* (pp. 480–488). New York, NY: Springer-Verlag.

Maslow, A. H. (1962). *Toward a psychology of being*. New York, NY: Van Nostrand.

Moss, C. J., & Poole, J. H. (1983). Relationships and social structure of African elephants. In R. A. Hinde (Ed.), *Primate social relationships: An integrated approach* (pp. 315–325). Oxford, United Kingdom: Blackwell.

Ogden, J. J., Finlay, T.W., & Maple, T.L. (1990). Gorilla adaptations to naturalistic environments. *Zoo Biology, 9*(2), 107–121.

Poole, J. H. (1994). Sex differences in the behavior of African elephants. In R. V. Short & E. Balaban (Eds.), *The differences between the sexes* (pp. 331–346). United Kingdom: Cambridge University Press.

Poole, J. H., & Moss, C. J. (2008). Elephant sociality and complexity: The scientific evidence. In C. Wemmer (Ed.), *Never forgetting: Elephants and ethics* (pp. 69-98). Baltimore, MD: Johns Hopkins University Press.

Shoshani, J., & Eisenberg, J. (1992). Intelligence and survival. In H. Shoshani (Ed.), *Elephants: Majestic creatures of the wild* (pp. 134-137). Singapore, Singapore: Weldon Owen.

Stoinski, T. S., Hoff, M. P., & Maple, T. L. (2001). Habitat use and structural preferences of captive lowland gorillas: The effect of environmental and social variables. *International Journal of Primatology, 22*, 431–447.

Viljoen, P. J. (1989). Spatial distribution and movements of elephants in the northern Namib desert region of the Kaokoveld, South West Africa/Namibia. *Journal of the Zoological Society of London, 219*, 1–19.

Wilson, M., Kelling, A., Poline, L., Bloomsmith, M. A., & Maple, T. L. (2003). Post occupancy evaluation of Zoo Atlanta's giant panda conservation center: Staff and visitor reactions. *Zoo Biology, 22*(4), 365–382.

Wilson, M. L., Bashaw, M. J., Fountain, K., Kieschnick, S., & Maple, T. L. (2006). Nocturnal behavior in a group of female African elephants. *Zoo Biology: Published in affiliation with the American Zoo and Aquarium Association, 25*(3), 173-186.

Wyat, J. R., & Eltringham, S. K. (1974). The daily activity of the elephant in the Rwenzori National Park, Uganda. *East African Wildlife Journal, 12*, 273–289.

www.ingramcontent.com/pod-product-compliance
Lightning Source LLC
Chambersburg PA
CBHW061502180526
45171CB00001B/9